A snow book, northern Scotland

Based mostly on the author's field observations in 1938–2011

Adam Watson

Publication of this book was aided by the generous sponsorship of Bert McIntosh of Crathes, Banchory and McIntosh Plant Hire (Aberdeen) Ltd, Birchmoss, Echt, Westhill, Aberdeenshire AB32 6XL, www.mphltd.co.uk

Published by Paragon Publishing, 4 North Street, Rothersthorpe, Northants, NN7 3JB, UK
First published 2011
© Adam Watson* 2011

All rights reserved. No part of this publication may be reproduced, stored in a retrieval system or transmitted in any form or by any means, electronic, mechanical, photocopying, recording or otherwise, without the prior written permission of the copyright owner.

*Clachnaben, Crathes, Banchory, Aberdeenshire AB31 5JE, Scotland, UK adamwatson@uwclub.net

ISBN 978-1-908341-12-9

Book design, layout and production management by Into Print
www.intoprint.net +44 (0) 1604 832149

Printed and bound in UK and USA by Lightning Source

Front cover photo. A confiding mountain hare stands in sunshine at the entrance to its daytime snow hole in soft powder snow on Carn Dubh in Glen Clunie near Braemar, 10 March 1970 (Adam Watson senior)

Contents

Acknowledgements 4
Introduction...4
Chapter
1. Early-summer snow-cover on Ben Macdui plateau, 1947–2010 5
2. Fresh lying summer snowfall on the Cairngorms, 1944–2010 16
3. Trend to later autumn snowfall on the Cairngorms, 1944–2010 29
4. Review of Scottish snow-bed survival, 1900–2010 37
5. Snow lasts till winter, north-east Scotland 1942–2010, west Scotland 1945–2010................55
6. Cairngorms glaciers in the 18th and 19th centuries highly unlikely 67
7. Vantage points for snow patches, from roads in north-east Scotland 72
8. Skiing in and near Aberdeen in the early 1950s..74
9. The remarkable snowstorm of early September 1976 75
10. Polygonal hollows and dirt on snow surfaces..85
11. Photographs showing the use of snow by hill birds and mammals......................................99
12. Photographs of some snow features and snow avalanches.. 115
13. Lichen and moss as indicators of snow-lie on cliffs, boulders, soil 124
14. Some photographs of snow mould on hill vegetation..133

Acknowledgements

Rhona Weir kindly gave me permission to reproduce a photo by Tom Weir. I thank photographers named in captions, who also gave permission. If no photographer is named in a caption, I took the photograph.

Introduction

In this book I record and illustrate some of my studies of snow patches and snow in Scotland over many years. This may be a useful basis for others studying snow patches and snow, now and in future. Because principles of snow and snow patches are general, the book may be of use to workers beyond Scotland. I hope it may interest the increasing numbers of walkers and mountaineers who appreciate snow and snow patches. It may also stimulate fresh observations that elucidate our understanding of snow and its effects.

Chapter 1. Changes in early-summer snow-cover on Ben Macdui plateau, 1947–2010

Summary

The proportion of ground covered by consolidated snow at the start of June was recorded for 64 years on Ben Macdui plateau in the Cairngorms massif of Scotland. It fluctuated greatly amongst years, with extremes of 2% and 98%. During the 64 years as a whole, it was correlated negatively with the calendar year, i.e. less cover as the years passed. In 1947–86 it showed no material trend in relation to the year, but in 1987–2010 was associated negatively with the year, being especially small in 2003–07, but slightly larger in 2008–09 and very large in 2010. It was correlated positively with the number of mornings when snow lay during the previous winter at Braemar, a village below the Cairngorms, and with the number of skier-days in the previous winter at Cairngorm ski area near Ben Macdui plateau and Glenshee ski area south of Braemar.

Introduction

This paper documents a 64-year run of data on snow-cover in the Cairngorms massif of Scotland. At the start of each June, I visually estimated the percentage of ground covered by consolidated snow on the main part of Ben Macdui plateau. This excluded ephemeral summer snowfalls that can give 100% cover briefly. Statistical analyses tested whether the data showed a trend over the years and whether the variability changed. Also I tested whether snow-cover was correlated with the previous winter's snow-lie at Braemar village and with skier numbers at nearby downhill ski-areas.

Methods

Study area and estimates of snow-cover

The area covered most of Macdui plateau as visible from the summit of Cairn Gorm and from the nearer view on a good vantage point at grid reference 992 021. It excluded ground north-east of that viewpoint, on the slopes south of Cairn Gorm.

At the start of June, I visually estimated the percentage of ground covered by deep consolidated snow from the previous winter and spring. The data excluded ephemeral summer snowfalls. An estimate required only a few seconds without fog, even during an otherwise foggy day. If 31 May had good visibility and the forecast indicated fog next day, I made the estimate on the late evening of 31 May. In almost all years I could do it on 1 June. On no occasion did fog last throughout 31 May–1 June. However, occasionally an ephemeral fresh snowfall persisted throughout both days. If so, I made the estimate on the next suitable day, usually 2 June, rarely 3 June. In such cases the winter's snow-pack did not lessen materially over 1–3 days, because the new snow and the cold air that precipitated it kept the winter's snow from melting further. In 1953, when I was in Canada, my father Adam Watson senior made the observation and took photographs. Rangers at Cairngorm Ski Area have sent photographs taken on request in two recent years and also now estimate the percentage snow-cover visually and almost daily in late May and early June.

As with visual estimates of anything, it is easy for the eye to detect a 1% difference at the low end and the high end of the scale (e.g. the difference between 0%, 1% and 2%, or 100%, 99% and 98%), but to find difficulty distinguishing a 5% difference in the middle of the scale (such as between 40 and 45%, or 55 and 60%. For this reason, estimates reflect this, such as using values of 2% and 98%, but in the middle parts of the scale I estimated only to the nearest 5% and often just to the nearest 10%.

Exceptional cover of deep consolidated snow, North Top of Ben Macdui at 2 km distance, viewed from the north on south side of Cairn Lochan, two walkers far left beside Lochan Buidhe, 21 May 1977

Cairn Lochan at 3 km distance, seen from the North Top of Ben Macdui, 21 May 1977

Braeriach from the upper Feith Buidhe, near horizon 1 km away, Braeriach summit 4 km away, 22 May 1977

South view for 1 km to the North Top of Ben Macdui, 3 June 1977 (Adam Watson senior)

A. Watson senior skis to Ben Macdui on 1 June 1977

A very snow-free spring, Cairn Lochan from north of North Top, 17 April 1981

North view from west of the summit of Ben Macdui, 17 April 1981

In some years I or rangers took photographs, which gave an independent record. The preceding photographs are examples.

Snow-lie at Braemar and Cairn Gorm, and skier numbers

For decades the Meteorological Office supervised a climatological station at Braemar village by the river Dee southeast of Ben Macdui, and summarised the data in publications (References) and now electronic files. In recent decades there was a climatological station at the foot of the ski area on Cairn Gorm. Observers at both stations noted the number of days when snow covered at least half of the ground at 0900 hours.

Scottish ski companies record the number of skier-days for each winter, based on tickets sold. I used data at Cairngorm ski area near Ben Macdui plateau and Glenshee ski area in the Mounth range south of Braemar. I did not use data from the Lecht ski area east of the Cairngorms, because large-scale snow-making equipment boosted skier numbers artificially there.

Statistical analyses

Because data on snow-lie in June comprised percentages, and all proportions are between 0 and 1, I transformed percentage values (angular transformation) before analysis.

Results

Whether snow-cover showed any trend as the years passed

Figure 1 shows data for snow-cover on Ben Macdui plateau at the start of June in 1947–2010. The graph in most cases shows marked ups and downs a year apart, a low value in one year being followed by a low one in the next year and then another high one. To this general picture there were four exceptions with runs of successive low or high values. For low values, a run of three occurred in 1948–50 and a run of seven in 2002–09. A run of successive high values lasted six years in 1965–70 and another for four years in 1977–80.

Statistical analyses showed snow-cover to be correlated negatively with the calendar year during the run as a whole, i.e. less as the years passed (Table 1). However, it was obvious from field experience and Fig. 1 that this overall decline resulted from a fall in recent decades. Because air temperatures at climatological stations show Britain to have been warmer since the late 1980s (Hulme & Jenkins 1998), this provided an objective criterion for splitting the time series. Hence I compared data in 1947–86 and 1987–2010.

In 1947–86, snow-cover showed no trend with the calendar year (Table 1), but in 1987–2010 was associated negatively with the year, i.e. tending to be less as the years passed. Up to 2008 the correlation coefficient was negative and statistically significant, but with data from the very snowy 2009 and 2010 included, the coefficient became weaker and this removed the significance (note below Table 1). Especially low in 2003–07, snow-cover rose in 2008 and 2009 and became very high in 2010. The long low trough in 2003–07 was an exceptional event in the entire run. The question arises whether the greater cover of 2008–10 indicates an end to the decline or a short-term rise before further decline. Observations in the next few years will tell.

Variation in snow-cover amongst years

Extremes in snow-cover were 2% and 98% (Table 2). The mean for 1987–2010 had a big variance. A useful index of dispersion that allows for differences in sample size is the variance divided by the mean. It was larger in the later years. I checked the validity of this difference with the F test for equal variances. The two sets did not have significantly unequal variances. In short, snow-cover in recent decades tended to be more variable than previously, but far from significantly so.

Snow-cover compared with snow-lie at Braemar and Cairngorm ski area

Snow-cover on Ben Macdui plateau at the start of June was correlated positively with the number of mornings in the previous winter when snow lay at Braemar and near the foot of Cairngorm Chairlift (Table 1). The number of mornings with snow at these latter two places was also positively correlated.

Snow-cover compared with skier numbers at downhill ski areas

Snow-cover on Ben Macdui plateau at the start of June was correlated positively with the number of skier-days during the previous winter at Cairngorm ski area. Likewise it was correlated positively with the number of skier-days at Glenshee ski area.

Discussion

In this paper I presnt evidence of a decline in snow-cover at the start of June on Ben Macdui plateau in recent years, one manifestation of a general decline in snow-across other Scottish hills. It coincides with a measured fall in the number of Scottish snow patches that survive until lasting winter snowfall. Patches surviving in recent years have been fewer and smaller than in earlier decades (Watson *et al.* 2006, 2007, 2010).

When winter precipitation is rain rather than snow, winter floods are more immediate and severe, and summer river-flows decline. The latter in turn dilutes water pollution less and creates poorer conditions for salmon and angling. If less summer snow-cover continues, effects on plants and animals in the alpine zone seem possible. There is now more doubt about this, however, because of snowier conditions in the last three years, especially in 2010.

Acknowledgements

I thank CairnGorm Mountain rangers Nic Bullivant, Attila Kish, Ruari MacDonald and Heather Morning for taking photographs on request in a few recent years, and Dr Julian C Mayes for reading an early version of the manuscript and giving valuable advice.

References

Hulme, M. & Jenkins, G. (1998). Climate change scenarios for the United Kingdom summary report. UK Climate Impacts Programme Technical Report No. 1. Climatic Research Unit, University of East Anglia, Norwich.
Meteorological Office (1962 and annually till 1991). Monthly weather report. Meteorological Office, Bracknell.
Watson, A., Duncan, D. & Pottie, J. (2006). Two Scottish snow patches survive until winter 2005/06. Weather 61, 132–134.
Watson, A., Duncan, D. & Pottie, J. (2007). No Scottish snow patch survives until winter 2006/07. Weather 62, 71–73.
Watson, A., Cameron, I., Duncan, D, & Pottie, J. (2010). Six Scottish snow patches survive until winter 2008/09. Weather 64, 184–186.

Figure 1. Percentage of ground covered by consolidated snow at the start of June on Ben Macdui plateau.

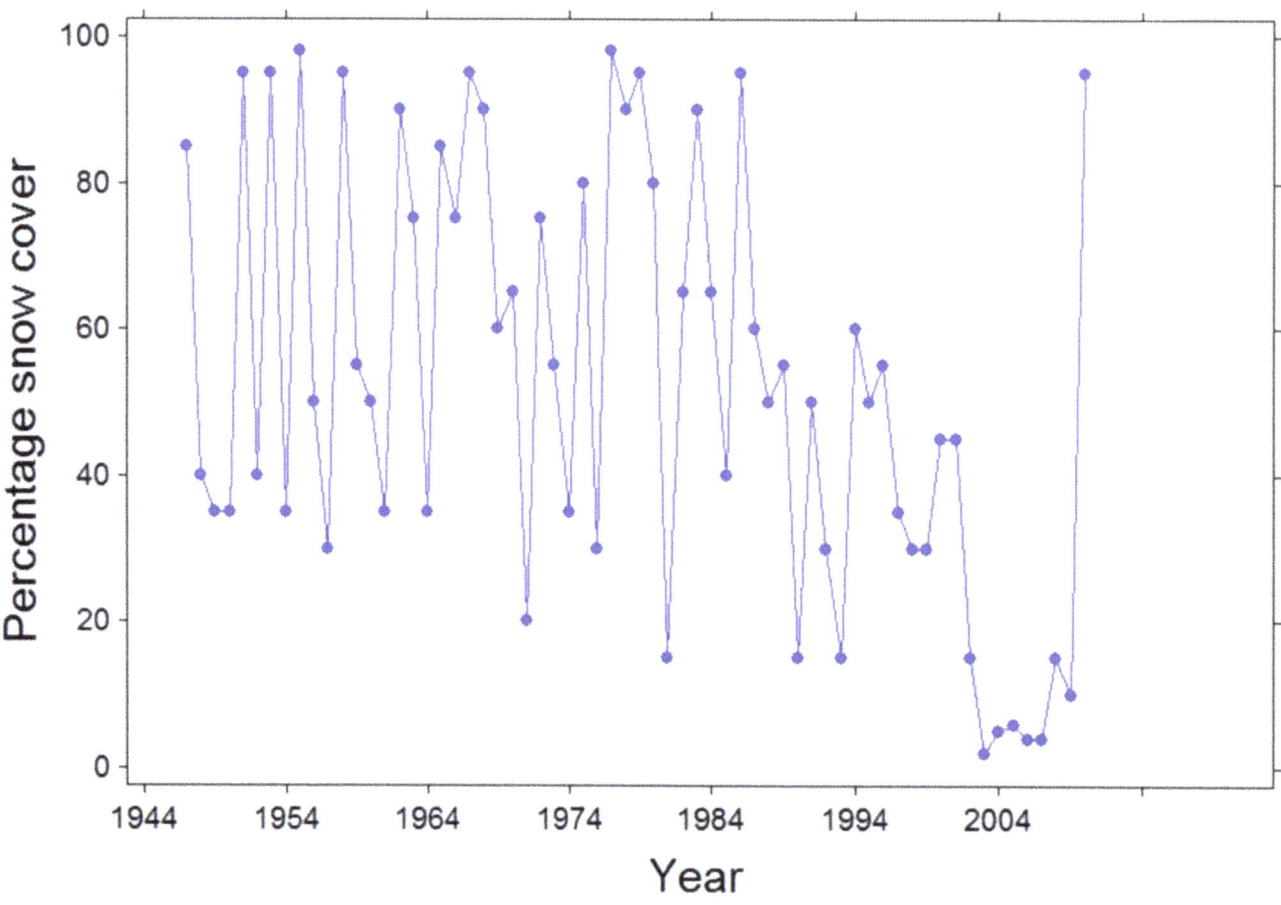

Table 1. Percentage snow-cover on Ben Macdui plateau at start of June compared with other factors, r being the correlation coefficient and P the probability of the result arising randomly.

Years	Compared with	n of years	r	P
1947–10	Calendar year	64	–0.447	0.0002
1947–86		40	0.098	0.5
1987–10		24	–0.371#	0.0738
1947–08	Braemar snow-mornings*	62	0.691	<0.0001
1947–86		40	0.633	<0.0001
1987–08		22	0.783	<0.0001
1984–93	Cairngorm snow-mornings*	10	0.712	0.021
1979–10	Cairngorm skier-days^	30	0.658	0.0001
1979–10	Glenshee skier-days^	32	0.650	0.0001

\# Value up to 2009 was –0.721, $P = 0.0002$, but inclusion of the very snowy 2010 reduced the coefficient and removed the significance.

* Number of days in the previous winter when snow lay at 0900 hours at Braemar (339 m, nearby at 327 m since 2005) and Cairngorm Chairlift bottom station (663 m). No Braemar data for 2006–2008, so nearby Balmoral used (The Snow Survey of Great Britain used them interchangeably as Balmoral/Braemar in Tables).

^ Number of downhill skier-days in the previous winter at Cairngorm ski area near Ben Macdui plateau and at Glenshee ski area near Braemar. The number of skier-days declined with the year, at Cairngorm ($n = 30, r = –0.805, P < 0.0001$) and Glenshee ($n = 32, r = –0.405, P = 0.021$). The decline was large since 1987, at Cairngorm ($n = 24, r = –0.795, P < 0.0001$) and Glenshee ($n = 24, r = –0.541, P = 0.0063$).

Like June snow-cover on the plateau, the number of Braemar snow-mornings was unrelated to the calendar year in 1947–86 ($n = 40, r = 0.15, P = 0.6$), but it declined in 1987–2008 ($n = 22, r = –0.484, P = 0.022$). With the much snowier 2009 and 2010 included, however, the correlation coefficient weakened and was no longer statistically significant ($n = 24, r = –0.358, P = 0.086$). The number of snow-mornings was positively correlated at Braemar and Cairngorm Chairlift ($n = 10, r = 0.936, P < 0.0001$).

Table 2. Sample size n, transformed* arithmetic mean percentage of snow-cover, transformed variance, index of dispersion (variance divided by mean) and untransformed minimum and maximum.

Years	n of years	Mean	Variance	Index	Minimum	Maximum
1947–86	40	55.2	303.6	5.50	15	98
1987–10	24	33.3	283.9	8.54	2	95

*Untransformed means in the two periods were 64.4 and 32.5.

A two-sample T test showed that the two means differed significantly ($T = 4.94, P < 0.0001$). Variances differed, but not significantly (folded F test for homogeneity of variances, $F = 1.07, P = 0.44$).

Chapter 2. Fresh lying summer snowfall on the Cairngorms, 1944–2010

Summary

Days of fresh lying snowfall were noted on the Cairngorms massif of Scotland in each of the months June–September during 1944–2010. The annual number of such snowfall-days was positively associated with the year in the first half of the run in 1944–76, though not significantly. During the second half in 1977–2010, it was negatively related to the year, almost statistically significant for June, and significantly for the summer total, i.e. fewer days with fresh lying snowfall as the years passed. The number of snowfall-days in each month and for all months varied more in the second half of the run than in the first, but not reaching statistical significance, and not at all in August and September. Summers with no snowfall-days were uncommon. Only one was recorded in the 46 years 1944–89, but four in the 20 years 1990–2010 during a period that included much-publicised climatic warming. This recent less snowy tendency may have become partly reversed in 2007–10, a period also with snowier winters.

Introduction

This paper documents a 67-year run of dates of fresh lying snowfall on the Cairngorms massif of Scotland. The same observer noted each year the dates when fresh snowfall lay. Analyses involved tests whether data showed statistically significant trends over the years, and whether variability between years changed.

Derry Cairngorm from Carn Crom after a snowfall with north gale, 22 June 1964

Ben Macdui from Derry Cairngorm, 22 June 1964

Study area and methods

The hills were on the north-east, east and south-east sides of the Cairngorms, especially Ben Avon and Beinn a' Bhuird. Because summer snowfalls came with northerly winds, these and other northern hills such as Cairn Gorm received heavier falls. Hills further south in the massif, such as Ben Macdui and Cairn Toul, and yet more so the hills further south than the Cairngorms, such as Lochnagar and Glas Maol, usually got less, often none. In these respects, snowfalls in summer and early-autumn were alike (Chapter 3). That chapter also describes where I lived, my vantage points, persons who assisted, and my checks with residents living in the Cairngorms when I suspected fresh snow but fog obscured the hills.

Data involve the number of days per month in June to September when a fresh snowfall during the day (or overnight since the previous day) lay on the Cairngorms at any altitude and any time. This differed from the Snow Survey of Great Britain, which involved observations whether snow covered at least half of the ground at 0900 hours. On some occasions, snow that fell and lay in late evening or early morning had gone by 0900 hours, but I recorded this as a snowfall day. Likewise I recorded daytime coatings that vanished by evening.

'Lying' meant covering most (more than half) of the ground or all ground. A fall on one day might leave snow for the next day, even if no new snow fell and lay in the next day. I counted this as two days. For brevity I used the term 'snowfall-day' to cover any day in a month when snow from a fresh snowfall that month was lying.

I excluded snowfalls (usually hailstones) during thunderstorms, because the aim was to record summer snowfalls in cold airstreams covering a wider region than the very local falls in thunderstorms. The latter often affected only part of one hill at a time.

For comparison I used skier numbers as an indication of the snowiness of the previous winter. Data are made publicly available by Scottish ski companies and also are published in summary form in newspapers. They involve the total number of skier-days in each winter skiing season from October through to May, based mainly on daily numbers of tickets sold and the company's estimates of the number of days used by the average person with a seasonal ticket.

Results

Number of snowfall-days amongst years

In 11 years I recorded fresh snow lying in three of the four months. In no year did I find it lying in all four months.

The 67-year run reveals differences within and among months and years (Fig. 1). It also shows the few summers when there were no days with fresh lying snowfall, a feature which has been more frequent since 1990. In the first half of the run, this occurred only in 1946 and 1967, but in the second half four times, in 1993, 1996, 2002 and 2006.

Table 1 shows mean dates and other data, including extreme monthly numbers of snowfall-days. Relatively few snowfall-days occurred in the mid 1940s and mid to late 50s, and particularly since 1992. They were frequent in the 1970s and 1985–87.

The longest run of Junes with no snowfall-days lasted five years in the early 1990s, and 1971–89 was notable for the large number of snowfall-days in June. July had fewer than June, and three long runs had none, up to 16 years in the 1980s and 1990s. August had the least snowfall-days of the four months, and four long runs of years had none, up to 17 years in 1988–2009. September was the summer month of most snowfall-days, with markedly high numbers in 1950, 1952, 1974–76 and 1986; and it might be argued that September was an autumn month rather than a summer one. The total annual number for all months June–September was particularly large in 1948, 1950, 1952, 1965, 1971–79, 1985–87, and 1991.

Number of snowfall-days in relation to calendar year

I compared each year's number of snowfall-days with the calendar year, to test for any trend over time. All correlation coefficients were positive in 1944–76 (Table 2). This suggested a trend towards more snowfall-days, but no coefficient was significant. During 1977–2010, in contrast, all coefficients were negative, almost significantly so for June and significantly with the total for all months. In short, there tended to be more snowfall-days as the years passed in 1944–76, but the opposite (and more strongly) in 1977–2010.

New snowfall, Cairn Lochan from Derry Cairngorm, 23 June 1957

Summers with no observed snowfall-days were uncommon, occurring only in 1946, 1993, 1996, 2002 and 2006. Thus one was observed in the 46 years 1944–89, but four in the 21 years 1990–2010 which included a period of well-publicised warming (2 x 2 tables, Yates' corrected $\chi^2 = 3.75$, $P = 0.0527$). The Poisson expression is a more commonly used method for studying infrequent events. Given the one case in 46 years, the expected annual frequency is 0.021, so only 0.441 cases would be expected in 1990–2010. The cumulative Poisson probability of observing four or more cases in 1990–2010 is about 0.0001. Hence the high frequency in recent years is very unlikely to be random.

There is one caution, involving a suggestion of more days of lying snowfall in the last few years. In June there have been larger values in the recent three years 2007–09 after a long run of years with low June values. Because of these larger June values, the totals for June–September have also been larger in the three years. If the last four years are excluded, the correlation coefficient for the total in June increases ($r = -0.533$, $P = 0.003$), and likewise that for the total in June–September ($r = -0.567$, $P = 0.0011$). On the other hand, there have been no July days with lying fresh snowfall since 2000, and no September days with it in 2008 and 2009. Future data from several successive years to come should help decide one way or the other.

Differences in means and variability between 1944–76 and 1977–2010

The index of dispersion, which indicates the extent of variability in the mean, was larger for June and very marginally for July in the second half of the run than in the first half, but for other months the opposite (Table 1). Analyses showed no statistically significant differences in the mean number of snowfall-days between the two halves of the run (Table 3). The *F* test revealed significantly unequal variances in the comparisons for June and August.

Snowfall-days on each day amongst years

Figure 2 shows a clumping on 2–7 June and a slighter one on 1–12 July. Sets of dates with no snowfall-days ran for four days from 16 July and from 22 July, and for seven and seven days from 7 and 21 August. Hence most of July and almost all of August were snow-free. Every September date amongst all years had one or more snowfall-days, with the second half of the month more snowy than the first half, and the last 10 days snowiest.

Wind directions during snowfall-days

Summer snowfalls came with northerly winds, often due north, and with one exception between north-west through north to north-east (Table 4). In that exception, a west-north-west wind brought snow in September. The wind had come from the north-west off Greenland, and then swung to west-north-west along the south side of a depression that was centred north-west of Scotland.

Snowfall-days and previous winter's skier-days at Cairn Gorm

During the 30-year run of years with data on skier numbers, since 1979, the number of summer snowfall-days on the Cairngorms massif was correlated positively with the total number of skier-days in the previous winter at the ski area on Cairn Gorm (Table 5). Hence, summers with high numbers of snowfall-days on the massif tended to follow snowy winters that had attracted many skiers to the only downhill ski area in the massif.

Discussion

The total number of snowfall-days in the whole summer was largest in 1973–79 and 1985–87, and for June in 1971–89. This might suggest generally cooler conditions in the 1970s and 1980s across northern Scotland and further north. In fact, Jones (1988) reported a cooling of mean annual land-based surface temperatures in 1967–86 across north-west Europe as far north as Iceland and all of Scandinavia, and including Scotland.

Figure 1 shows that unusually few days of fresh summer snowfall occurred in 1944–46 and again since 1992 than in the intervening decades. This recent less snowy tendency may have reversed to some extent in 2007–10. That has been a period with snowier winters and also more summer snow patches surviving until lasting winter snow (Watson *et al.* 2009, 2010).

Future work could involve a comparison of snowfall dates with published data on air temperatures, depressions and

New snow, Coire an Lochain of Cairn Gorm, 10 July 1971 (Adam Watson senior)

fronts. Another study could involve comparison of the number of summer snowfall-days with the number of days of snow-lie in the previous winter as published in the Snow Survey of Great Britain.

Acknowledgements

Neil Baxter, Colin Bruce, Nic Bullivant, Stewart Cumming, John Duff, David Duncan, Stuart Gordon, James Grant, Bob Kinnaird, Gus Jones, Peter Holden, Attila Kish, Ruari MacDonald, Heather Morning, Ian Murray, Dave Patterson, Tom Paul, Eric Pirie, John Pottie, George Reid, John Robertson, Sandy Walker, Tim Whittome, and the late Bob Clyde, Fred Harper, Eric Langmuir, Davy Rose and Bob Scott checked in the Cairngorms, on request. I am grateful to Dr Julian C Mayes for reading the manuscript and giving valuable advice.

References

Jones, P.D. (1987). Hemispheric surface air temperature variation: recent trends and an update to 1987. Journal of Climate 1, 654–660.

Watson, A., Cameron, I., Duncan, D. & Pottie, J. (2010). Six Scottish snow patches survive until winter 2009–10. Weather 65, 196–198.

Watson, A., Duncan, D., Cameron, I. & Pottie, J. (2009). Twelve Scottish snow patches survive until winter 2008/09. Weather 64, 184–186.

Watson, A., Pottie, J. & Duncan, D. (1999). Only one UK snow patch lasts until winter 1998/99. Weather 54, 369–374.

Watson, A., Pottie, J. & Duncan, D. (2000). Six UK snow patches last until winter 1999/2000. Weather 55, 286–290.

Watson, A., Pottie, J. & Duncan, D. (2001). Forty-one UK snow patches last until winter 2000/01. Weather 56, 404–407.

Table 1. Sample size n, arithmetic mean, variance, index of dispersion (variance divided by mean), and minimum and maximum number of days of snowfall that lay.

	Years	n of years	Mean	Variance	Index	Minimum	Maximum
June	1944–76	33	1.30	2.41	1.85	0	4
	1977–10	34	2.15	5.71	2.66	0	9
July	1944–76	33	0.52	1.57	3.02	0	5
	1977–10	34	0.41	1.58	3.85	0	6
August	1944–76	33	0.27	0.45	1.67	0	3
	1977–10	34	0.12	0.10	0.83	0	1
September	1944–76	33	3.15	6.70	2.13	0	10
	1977–10	34	2.12	4.41	2.08	0	8
Total	1944–76	33	5.24	12.56	2.40	0	13
	1977–10	34	4.79	13.20	2.76	0	15

New snow, Cairn Lochan from Cairn Gorm, 18 June 1971 (Adam Watson senior)

Table 2. The number of days of lying snowfall in relation to calendar year; r is the correlation coefficient and P the probability of the observed result arising randomly.

		r	P
June	1944–76	0.335	0.056
July		0.023	0.898
August		0.245	0.170
September		0.229	0.201
Total		0.368	0.035
June	1977–2010	−0.314	0.071
July		−0.266	0.128
August		−0.009	0.958
September		−0.117	0.546
Total		−0.340	0.037

Table 3. Two-sample T tests of differences in the mean annual number of days of snowfall that lay in 1944–76 compared with 1977–2010, and folded F tests of the homogeneity of variances.

	T	P (for T)	F	P (for F)
June	−1.71	0.09	2.37	0.008
July	0.34	0.74	1.01	0.49
August	1.19	0.09	4.25	<0.0001
September	1.8	0.077	1.52	0.12
Total	0.51	0.61	1.05	0.45

Comparisons of early v. later years revealed no statistically significant difference in means and two significant differences in equality of variances. When data for the more snowy recent years 2006–10 are excluded, the comparison for September shows a bigger difference ($T = 2.20$, $P = 0.048$, $F = 1.83$, P (for F) = 0.052), but even so, just barely statistically significant.

New snow on Cairn Lochan, 18 June 1971 (Adam Watson senior)

Table 4. Sample size *n*, and wind directions on days when fresh snowfall lay, showing for each wind direction the percentage for each month.

	n of snowfall-days	WNW	NW	NNW	N	NNE	NE
June	116	0	1	2	90	5	3
July	31	0	0	0	100	0	0
August	12	0	0	0	100	0	0
September	176	1	14	16	62	7	2
Total	335	0.3	8	9	81	5	2

These are rough approximations, because direction sometimes changed during a day. Directions shown are for early morning on each snowfall-day.

Table 5. The number of snowfall-days in each month and in total for the summer, compared with the number of skier-days in the previous winter (October–May) at the downhill ski area on Cairn Gorm in 1979–2010 (n = 30 years of data on skiers), with r the correlation coefficient and P the probability.

	June	July	August	September	Total
r	0.393	−0.181	0.213	0.390	0.480
P	0.032	0.34	0.259	0.033	0.007

The number of skier-days per winter varied from 44 800 in 2004 to 391 000 in 1988. During 1979–2010 it declined with the year (r = −0.805, P < 0.0001), as did the number of summer snowfall-days (r = −0.309, P = 0.095).

New snow, Ben Macdui from Cairn Lochan, less snow southwards and hardly any on Cairn Toul further right, 18 June 1971 (Adam Watson senior)

Figure 1. Number of summer days when fresh snow lay on the Cairngorms during each year.

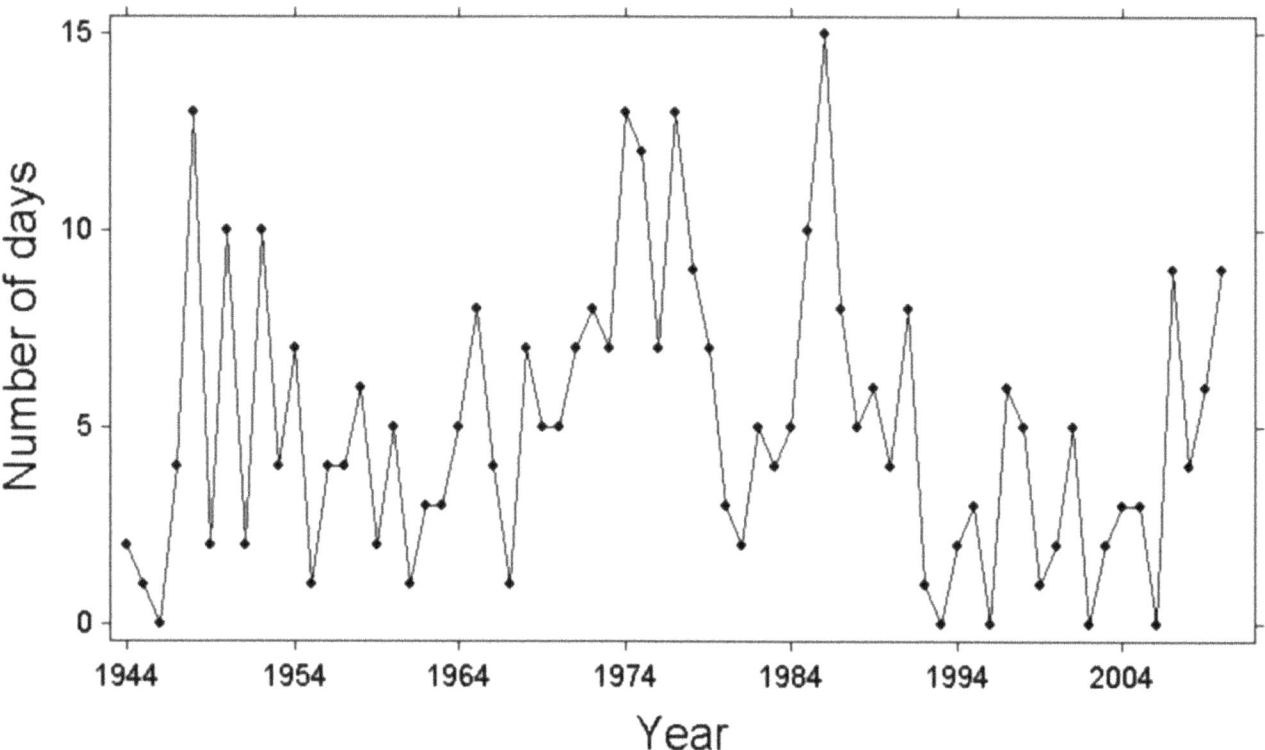

Figure 2. Number of days when fresh snow lay on the Cairngorms during each successive date, in all years combined.

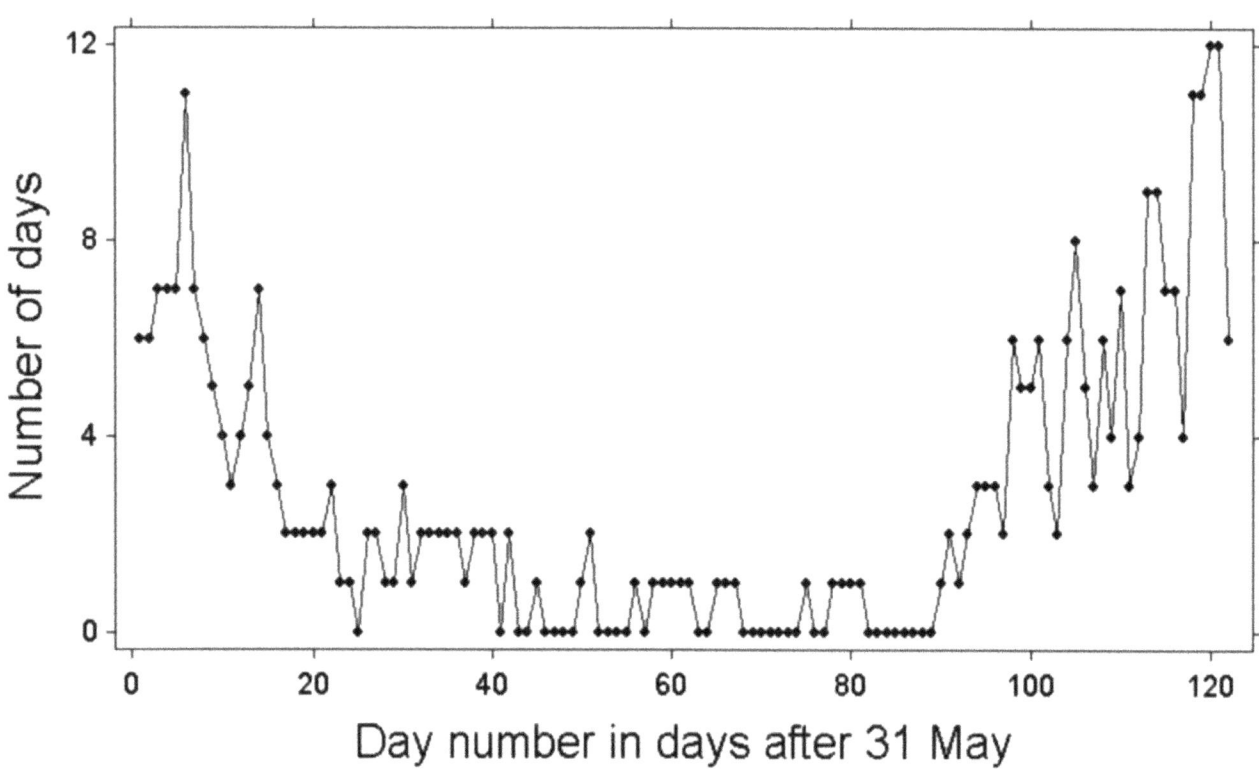

Chapter 3. Trend to later autumn snowfall on the Cairngorms, 1944–2010

Summary

The dates of the first autumn snowfall lying on the Cairngorms massif were noted in 1944–2010, and also the dates of the first snow that lay until lasting winter snow. During the 67-year run as a whole, each kind of date was positively and significantly correlated with the year, i.e. later as the years passed. In the first half of the run, each kind showed very weak insignificant positive associations with the year. In the second half of the run, both kinds were positively associated with the year, but only the relationship with the date of first snow came close to being significant. Because all correlation coefficients were low, most of the variation in dates was due to factors other than the calendar year. Both kinds of date became more variable in the second half of the run, with greater extremes. There was a suggestion of earlier dates in the last three years, but whether this indicates a cooler tendency will become clear only with a longer run of years.

Introduction

The main evidence for climatic change in recent decades was direct, involving surface air and sea temperatures (e.g. Harrison 1996; Hulme & Jenkins 1998). Indirect changes are useful for confirmation and prediction, such as the state of glaciers and snow patches. Another indirect change is the earliest date of autumn snowfall. The present paper documents a 67-year run of such dates on the Cairngorms massif of Scotland. The same observer noted each year the first date of fresh snowfall after the end of August, and the first date of fresh snowfall that remained until the coming of lasting winter snow. I tested whether the data showed any trends and whether variability changed between years.

A well known run is in the annual reports of the Snow Survey of Great Britain (Hawke & Champion 1952, 1953, 1954, and Meteorological Office annually in 1954–91). An eight-year run is in an old diary (Earl of Fife 1783–90; Green 1970; Watson 1983). I compared dates in my run with these others.

Study area

The chief hills seen were on the north-east, east and south-east sides of the Cairngorms, especially Ben Avon and Beinn a' Bhuird. Because autumn snowfalls came with winds from north-west through north to north-east, the named two hills and the other high northerly ones of Cairn Gorm, Braeriach, Beinn Mheadhoin and Beinn a' Chaorainn usually received the heaviest falls. High hills further south in the Cairngorms, such as Ben Macdui and Cairn Toul usually got far less snow, often none. This applied yet more strongly on high hills yet further south, such as Lochnagar and Glas Maol.

In the 1700s the Earl of Fife kept a diary of weather during autumn deer-hunting at Mar Lodge on the Dee side of the Cairngorms. He saw Beinn Mheadhoin, Beinn a' Chaorainn and Beinn a' Bhuird daily, as well as more southerly hills in the Cairngorms.

First dates of autumn snowfall in the Snow Survey refer to various hills in the west, north and central Highlands as well as the Cairngorms, and in many years the notes do not mention locations. I therefore analysed only those Survey data where the location was given as the Cairngorms. It should be noted that many Survey dates for the Cairngorms involved the Spey or north side of the massif, unlike the Earl's from Deeside and also mine from Deeside and other places east of the massif.

Ben Macdui from Carn Crom after new snowfall from north, 21 September 1963

Methods

Over the years I found fresh snowfalls lying on the Cairngorms in all high-summer months June, July and August. On average, August was the month with fewest days of fresh snow lying, so I decided to use 31 August as a starting point, and for analysis I converted each date of fresh lying snow to the number of days after 31 August. I made two small adjustments to this. When all years were combined, the longest run of summer dates without new snow-lie was 22–28 August, and thereafter every date had snow-lie. I therefore continued to use days after 31 August as a starting point, but allocated −2, −1 and 0 for new snow-lie on 29, 30 and 31 August.

Each autumn I noted the first date when I observed fresh snow lying on the Cairngorms at any altitude, even though all of it melted within a day. Also I recorded the first date of fresh snow that lay until lasting snow. I define lasting snow as snow that persisted through the winter until the following spring, at least at one site holding a persistent snow-bed. As a minimum, I regard lasting snow as a covering on part or all of the upper surfaces of old snow from the previous winter, visually obvious through binoculars from viewpoints 2 km or more distant. At the most, it was a near-complete covering of the ground.

For the sake of brevity below, I call these two dates the first date of snow-lie and the first date of lasting snow-lie. In some years the two coincided. In many years, however the first snow-lie vanished and lasting snow came later, in some cases weeks or even months later, after successive new snow-lie had all melted.

I excluded ephemeral falls (usually of hail) during thunderstorms in warm weather, because these were highly localised and often affected only one hill or part of a hill. The aim was to record fresh snowfalls in cold airstreams, when snow fell on most or all hills in the Cairngorms and often more widely.

Although a fresh snowfall was conspicuous to the naked eye, I used binoculars. They were useful for checking in detail whether an earlier snowfall had vanished or not from the main hollows. I recorded the lowest altitude to which snow lay in the morning. Often it melted quickly up to a higher altitude within the morning, and sometimes all of it disappeared within a few hours, whereas at the other extreme a heavy fall with drifting continued to whiten the hills for weeks. I noted the fall as heavy, light, or a dusting, with drifting or without, the direction of drifting, the aspect of drifts, and estimated depths of drifts. I do not present this detail below, but it was a useful discipline for fieldwork, especially for utilising brief good visibility on days of bad weather. Because of fog, haze or showers, hills might be visible for only a brief glimpse.

In 1944–47 I stayed mainly at Turriff south of Banff, but could see the east Cairngorms from north-east of the town and was often in the Cairngorms at weekends. During 1948–52 I worked in the Cairngorms during early autumn, and although I was at Aberdeen in late autumn 1948–51, hills beside the city afforded views of the Cairngorms. Through late autumn 1952 and early autumn 1953 I was in Canada, but my father made observations from near Turriff and on frequent weekends in the Cairngorms. During late autumn 1953 and autumn 1954–57 I again lived in Aberdeen, but spent most weekends and many weekdays in the Cairngorms where I did fieldwork on ptarmigan for a PhD. In autumn 1958–60 I lived in Glen Esk, but saw Lochnagar almost daily, and often saw the Cairngorms on days while doing fieldwork on red grouse in upper Deeside and the east Cairngorms. My father observed fresh snow on many weekends and weekdays in 1954–60, as did Patrick D. Baird in 1954–59 when he studied a snow-bed near the summit of Ben Macdui beside a weather station that he erected there (Baird 1957). In 1961–2010 I stayed near Banchory in lower Deeside, where on clear days I saw the Cairngorms from near my house.

Hence in most autumn weeks I did not live in the Cairngorms, but frequently saw them from vantage points. If I suspected new snow but fog obscured the hills, I telephoned friends resident in the hills. Usually they had seen the hills briefly and could recount details of snow cover, altitude and aspect. This made them used to looking for snow. However, the study could not have been done had I been living outside north-east Scotland or going elsewhere on holiday in autumn.

Results

Variation in first dates amongst years

Figure 1 illustrates the 67-year run. The earliest date for first lying snow was 28 August, by coincidence in 2010 which was the last year in the run, the latest on 27 October (Table 1). Likewise the earliest date for the first lying snow that stayed till lasting winter snow was 5 September 1976, whereas the latest came on 5 December 1983. The 1976 case involved heavy snow with severe drifting. In the 1983 case, several earlier snowfalls had vanished in subsequent mild weather.

Whether first dates became earlier as the years progressed

During the run as a whole, both kinds of date were positively correlated with the year (Table 2). Hence the trend was for first dates of lying snowfall and of lasting snowfall to come later as the years passed. Although both correlations were significant, coefficients were very small, accounting for little of the variation in dates. Hence other factors must cause most of the variation, doubtless random weather.

A second test was to compare the first and second halves of the run. I used two broadly equal halves, 33 years in 1944–76 and 34 in 1977–2010. Dates in the first half of the run showed almost no association with the year. In the second half they were positively associated with the year, almost significant for dates of first snow-lie, not for dates of lasting snow.

A third test was to compare mean dates in the first half of the run with those in the second half. Analyses showed that the second half had later dates than the first half (Tables 1 and 3). Differences between the two means were strongly significant for first snow-lie, just significant for lasting snow-lie.

Because there has been a warmer climate in years since 1990 (Hulme & Jenkins 1998), I compared the mean dates in the recent 1990–2010 set with those in 1944–89. In this analysis, differences between means were again strongly significant for first snow-lie, but not at all for lasting snow-lie.

The first date came unusually late in 2006, but earlier in the next three years, while the lasting dates in 2008 and 2009 also arrived earlier after a run of four years with late dates. Whether or not this indicates a cooler recent tendency will become evident only with data from a longer run of years.

Whether first dates have become more variable

Dates in recent years appeared to show greater variation, with more extremes, than in earlier years (Fig. 1). Statistical analysis revealed that data for 1990–2010 did have larger variances (Table 1). This applied also to the second half of the run as compared with the first half, even though in this instance the sample sizes for number of years were almost identical, and thus an equal variance would be expected on the basis of sample size.

A useful index of dispersion that allows for differences in sample size is the variance divided by the mean. It also was larger in the later sets of years. I checked its validity with the *F* test for equal variances (Table 3). This supported the assumption that the two sets had unequal variances, significantly so in two cases. In short, dates in recent decades have tended to be earlier, and more variable.

First dates of fresh lying snowfall and of lasting snowfall compared

I compared both dates in the same year, to find whether they were associated. The analysis showed almost no association ($n = 66$ years, $r = 0.10$, $P = 0.4$). Associations were also very weak when compared within the first half of the run, within the second half, and within the 1944–89 and the 1990–2010 periods. Hence, for example, an early date of lying snow was not followed by an early date of lasting snow.

Number of days between first dates and dates of lasting snowfall

Within years, this number varied from 0 in several years when the two coincided, to 90 days in 1994 (Table 1). Neither number showed a correlation with the calendar year (Table 2). The mean number of days did not differ significantly between the first half of the run and the second half. However, the number of days showed a greater variance in the second half, and a test showed significantly unequal variances (Table 3).

First dates of fresh lying snowfall compared with other data

The mean date of the first lying snowfall in the 1944–2010 run resembled that in the Earl's years of 1783–90. Furthermore, first dates in the 1944–2010 run were correlated positively with those in the same years of the Survey ($n = 11$ years, $r = 0.759$, $P = 0.007$).

Discussion

Because the Earl concentrated almost wholly on deer-hunting, his weather notes are not strictly comparable with my 1944–2010 data. He went to different hills on successive days and occasionally took a day off for activities in woodland or riverbank, so he did not see the highest tops daily. Hence he may well have overlooked a light ephemeral dusting of snow there, or even a heavier morning snow-cover that vanished later in the day.

The Snow Survey is also not strictly comparable with my 1944–2010 data. Some Survey observers lived outside the Cairngorms, where conditions often differ. For instance, the Survey recorded the snowline on the summit of Ben Nevis throughout September 1953, whereas in the Cairngorms the weather was dry on 1–2 September and the first snow did not fall until 3 September. Here are two recent instances from 1999 and 2000. The first lying snowfall in 1999 lasted till winter in the Cairngorms, but vanished in the west, where lasting snow did not come until a month later (Watson *et al.* 2000). In 2000 by contrast, the first snow came earlier in the west because dry weather prevailed in the Cairngorms (Watson *et al.* 2001). For these reasons, in the Results above I used for analysis only those years when the Survey report mentioned the Cairngorms specifically in relation to a given first date.

Furthermore, Survey observers were instructed to record snow-lie at 0900 hours if snow covered half or more of the ground. Hence they would not record an ephemeral light fall at an earlier hour if it had vanished by 0900, or at a later hour if it vanished by next morning. Also, some observers viewed the north side of the Cairngorms. The north or north-west winds that typify autumn snowfalls often drift snow into the main snow-bed hollows on the south or east sides. These may hold snow that lasts until winter, even though hollows on the north side hold little or none after the first day. This contrast occurs frequently, e.g. in 1998 (Watson *et al.* 1999).

Acknowledgements

I thank Neil Baxter, Colin Bruce, Nic Bullivant, Stewart Cumming, John Duff, David Duncan, Stuart Gordon, James Grant, Bob Kinnaird, Gus Jones, Peter Holden, Attila Kish, Ruari MacDonald, Heather Morning, Ian Murray, Dave Patterson, Tom Paul, Eric Pirie, John Pottie, George Reid, John Robertson, Sandy Walker, Tim Whittome, and the late Bob Clyde, Fred Harper, Eric Langmuir, Davy Rose and Bob Scott for checking in the Cairngorms, on request, and Dr Julian C Mayes for reading the manuscript and giving valuable advice.

References

Baird, P.D. (1957). Weather and snow on Ben MacDhui. Cairngorm Club Journal 17, 147–149.

Earl of Fife (1783–92). Journal of the weather at Marr Lodge. Handwritten journal, now in Special Collections, King's College, University of Aberdeen.

Green, F.H.W. (1970). Weather notes from an Aberdeenshire diary, 1783 to 1792. Weather 25, 553–554.

Harrison, J. (1996). Changes in the Scottish climate. Botanical Journal of Scotland 49, 287–300.

Hawke, E.L. & Champion, D.L. (1952). Report on the snow survey of Great Britain for the season 1950–51. Journal of Glaciology 2, 25–37.

Hawke, E.L. & Champion, D.L. (1953). Report on the snow survey of Great Britain for the season 1951–52. Journal of Glaciology 2, 219–228.

Hawke, E.L. & Champion, D.L. (1954). Report on the snow survey of Great Britain for the season 1952–53. Journal of Glaciology 2, 356–362.

Hulme, M. & Jenkins, G. (1998). Climate change scenarios for the United Kingdom. *Summary report*. UK Climate Impacts Programme Technical Report No. 1. Climatic Research Unit, University of East Anglia, Norwich.
Meteorological Office (1954). Snow Survey of Great Britain, season 1953–54. Meteorological Magazine 83, 353–359.
Meteorological Office (1955). Snow Survey of Great Britain, season 1954–55. Meteorological Magazine 84, 361–366.
Meteorological Office (1956). Snow Survey of Great Britain, season 1955–56. Meteorological Magazine 85, 353–361.
Meteorological Office (1957 and annually till 1991). Snow Survey of Great Britain. Meteorological Office, Bracknell.
Watson, A. (1983). Eighteenth century deer numbers and pine regeneration near Braemar, Scotland. Biological Conservation 25, 289–305.
Watson, A., Pottie, J. & Duncan, D. (1999). Only one UK snow patch lasts until winter 1998/99. Weather 54, 369–374.
Watson, A., Pottie, J. & Duncan, D. (2000). Six UK snow patches last until winter 1999/2000. Weather 55, 286–290.
Watson, A., Pottie, J. & Duncan, D. (2001). Forty-one UK snow patches last until winter 2000/01. Weather 56, 404–407.

figure 1. The dates of the first lying snow and of the first snow that lasted till winter.

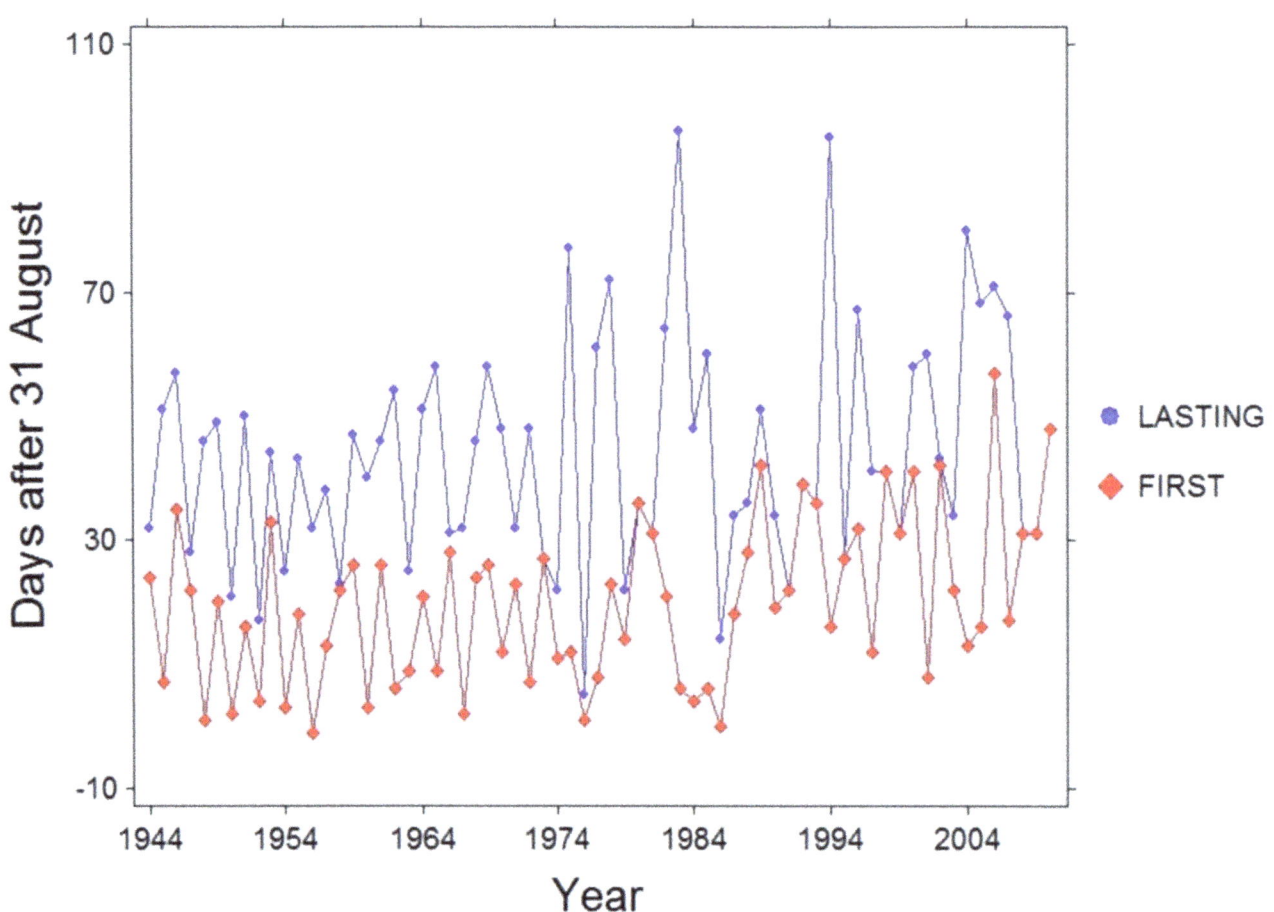

Only the red dot shows when both dates coincide.

Table 1. Sample size *n* (number of years), arithmetic mean, variance, index of dispersion (variance divided by mean), and minimum and maximum dates of the first snowfall that lay and the first that lay until lasting winter snow.

Dates	Years	n	Mean	Variance	Index	Min	Max
First	Earl of Fife	8	20.5	152.0	7.4	0	34
	Snow Survey	11	14.9	109.7	7.4	−1	35
	1944–76	33	14.9	106.7	7.2	−1	35
	1977–10	34	23.1	199.2	8.6	−3	57
	1944–89	46	15.5	127.5	8.2	−1	42
	1990–10	21	26.2	197.0	7.5	−3	57
Lasting	1944–76	33	39.5	220.6	5.6	5	77
	1977–10	34	48.5	420.3	8.7	14	96
	1944–89	46	41.9	306.6	7.3	5	96
	1990–10	21	48.7	390.2	8.0	22	95
First to lasting	1944–76	33	23.5	272.9	11.6	0	65
	1977–10	34	21.1	666.1	31.5	0	90

Table 2. Dates (defined in Table 1) compared with calendar year, *r* the correlation coefficient and *P* the probability.

Dates	Years	n years	r	P
First	1944–10	67	0.321	0.008
	1944–76	33	−0.055	0.76
	1977–10	34	0.204	0.25
Lasting	1944–10	67	0.228	0.063
	1944–76	33	−0.002	0.99
	1977–10	34	0.054	0.76
First to lasting	1944–10	67	−0.123	0.32
	1944–76	33	−0.011	0.95
	1977–10	34	−0.241	0.17

Table 3. Mean dates of first and lasting snowfalls compared in early with later years, using two-sample T tests with folded F tests for homogeneity of variances.

Snowfall	Comparison	T	P (for T)	F	P (for F)
First	1944–76 v. 1977–10	−2.72	0.0085	1.82	0.047
	1944–89 v. 1990–10	−3.22	0.002	1.54	0.11
	1783–90 v. 1990–10	−1.01	0.32	1.30	0.38
Lasting	1944–76 v. 1977–10	−2.06	0.044	1.91	0.036
	1944–89 v. 1990–10	−1.35	0.19	1.27	0.25
First to lasting	1944–76 v. 1977–10	0.45	0.66	2.44	0.007

Chapter 4. Review of Scottish snow-bed survival, 1900–2010

Summary

I review data on Scotland's most persistent snow at Garbh Choire Mor in the Cairngorms, and on patches that survived till winter at other sites in north-east Scotland and in west Scotland. Next I summarise earlier reviews of Scottish snow survival, and lastly test for trends. The data show statistically significant declines in the persistence of snow in comparison with years up to the early 1930s, especially since 1990. Snow patches have survived more often in the 1960s–80s than in the few decades before or since.

Introduction

This review is in four parts. First, I document observations on Scotland's most persistent snow at Garbh Choire Mor on Braeriach. Secondly I review other authors' published information on snow that lasted till winter in north-east Scotland, and separately in west Scotland. Thirdly there is a summary of earlier reviews of Scottish snow survival, including notes on errors. Fourthly I collate data from all sources, to test for trends.

I use the terms survived and survival to mean old snow that remained until the first lasting snowfall of the next winter (not the first snowfall, which was often ephemeral), For brevity I use AW for my own name. Observations by others were personal communications unless stated otherwise.

Study areas and methods

These are described in Watson *et al.* (1994, review 2002). Most data could not be normalised, so I used non-parametric tests for statistical analyses.

Results

Data of AW and colleagues on UK's most persistent snow at Garbh Choire Mor

Data are in Table 1, and in Table 2 of the review by Watson *et al.* (2002). The section below identifies a few snow patches by the names given by rock climbers to their routes up the cliffs above the snow. The climbing routes were sketched in Nisbet *et al.* 2007; see also Table 2 in Watson *et al.* 2002). The annual number of patches surviving varied considerably (median 2, mean 2.85, range 0–22, $n = 73$ years), as did their total length (median 36, mean 62.15, range 0–420, $n = 69$). The annual number was related to the total length in the same year ($r_s = 0.910$, $n = 69$, $P <0.0001$). In 10 years only one patch survived; in nine of these years it was below Sphinx Ridge and in one year (1953) below Pinnacles Buttress. The Sphinx snow survived all years in 1934–2010 except 1953, 1959, 1996, 2003, and 2006. In early autumn it covered a smaller area than the patch under Pinnacles Buttress, but in all years except 1953 it was deeper, as Gordon (1925) had noted.

Two patches on 1 September 1945 (AW) declined to one in mid September (Perry 1948), and old snow 17.5 x 12 ft x c30 inches deep lay at the upper hollow (i.e. below Sphinx) on 20 November when new snow had fallen (S. Gordon letter to AW, and Gordon 1951). In 1949, AW saw two small patches on 26 September, but on 16 October only one remained, 9 ft long (S. Gordon, letter to AW, reporting a Braemar man's visit). There was frost overnight on 15/16th and lasting snowfall overnight on 18/19th (AW), so the snow would have survived. This was its smallest extent, apart from years when all snow vanished. Only the Pinnacles patch remained on 27 September 1953 (AW senior), but it survived (see photograph of it taken on 17 October 1953 by AW senior, in Richardson 1954). In 1954, two patches remained on 24 September, the Pinnacles one very small (M. Smith). The Pinnacles one vanished later, but the Sphinx patch was 20 ft long on 31 October when new snow covered all but 12 square feet of the old snow (AW and P.D. Baird).

On 22 July 1959, A. Tewnion (letter to AW) photographed the already small patches and thought they would vanish. M. Smith measured the Sphinx snow on 31 August 1959 as 11 x 8 x 2 ft. After hot days it had gone by 8 September (P.D. Baird on Ben Macdui, told to AW), as Gordon (1963) confirmed at the site on the 10th, one of a succession of hot days. I judge that it would have vanished by 5 September. On the 13th, a sunny day, A. Watson senior, (diary) noted the warmth in the Pinnacles bowl at 1430 GMT as remarkable, but slightly cooler at Sphinx. The depth of each bowl surprised him, for in his previous experience the snow's fairly flat upper surface had hidden the bowl. A. Tewnion visited the snow-less bowls on 7 October 1959. The Sphinx snow had occupied a scoop 20 ft across, 15 ft out from the cliff-foot, and D. Duncan in 2003 estimated the greatest depth of the Sphinx and Pinnacles bowls as about 60 cm. A. Tewnion reported the snow's disappearance and illustrated it with a photograph (*Aberdeen Press & Journal*, 16 October 1959).

On 27 September 1961, Tewnion noted that ice 3-8 inches wide formed the rim of each patch, and the snow below a depth of three inches was very hard. On 26 September 1964 he measured the Pinnacles snow as 35 x 9 x 2 ft, and Sphinx 24 x 21 x 4 ft.

Smith (1961) rated Sphinx Ridge, first climbed in May 1952, as 'Very Difficult up to July'. It 'is something of a curio; for as the snow gradually recedes from its base (average year build-up estimated 80 feet at peak) so a smooth nose is brought to light which has so far prevented ascent of the ridge in the latter months of the year. This nose will usually make its appearance by the end of July'. The triangular face at the foot, tapering to a nose-like point, has since been climbed (graded as Severe by Fyffe & Nisbet 1995). Smith (1961) reported that snow depth under the main buttresses was 'estimated at 100 feet in a heavy year in one case' (told to him by AW), that the cornice reaches 35 feet high, and crevasses and bergschrunds occur. Watson (1998) wrote, 'A 30 m vertical depth of snow occurs in some springs', and his photographs on 19 April 1951 after a long snowy winter show that snow 'almost buried the rock climbs south of Sphinx Ridge'. On 27 September 1961, A. Tewnion (letter to AW) found a bergschrund 6 feet deep between the top of the Sphinx patch and the cliff.

Pottie (1995) gave scores for snow extent, using photographs taken about 5 October annually in 1975–93. Although much melting occurs after that date in some years, his scores were correlated with AW's data on the number of surviving patches in the same year ($r_s = 0.655$, $n = 19$ years, $P = 0.0029$) and their total length ($r_s = 0.773$, $P = 0.0002$).

Watson *et al.* (2002) reviewed the annual number and total length surviving in 1971–2000, and Watson *et al.* (1997–2011) gave detailed accounts annually for 1996–2010. Disappearance loomed on 31 August 1976 with only two patches left (AW), but a deep snowfall in early September (Hudson 1977, Pottie 1979) covered them and lasted until winter. All snow vanished there by 23 October 1996, 23 August 2003, and (estimated) by 1–2 October 2006.

Cliffs shade the snow in the afternoon and evening. Lying further east and lower, the Pinnacles snow receives sunshine longer than at Sphinx. The narrow Pinnacles Buttress backs it to the west-north-west, with the fairly wide Pinnacle Gully between the buttress foot and Sphinx Ridge's foot, whereas a longer continuous cliff backs the Sphinx snow, with Sphinx Ridge rising to the west-south-west. Gordon (1925) wrote that the snow 'even at midsummer when the sun is highest is in shadow shortly after midday', but here he overstated. Full sunshine bathed both patches at 1230 GMT on 16 July 2003 (D. Duncan told to AW). The Sphinx snow lay in shade at 1200 on 15 September 2002, and the Pinnacles snow in sunshine, whereas by 1230 a small part of the Pinnacles snow came into shade (J. Irvine told to AW). On 9 October 2004 the snow at Sphinx fell into shade at 1130, whereas at Pinnacles it stayed in sunshine till 45 minutes later (DD).

Others' published information for north-east Scotland

On 18 July 1904, Johnstone (1905) wrote of Fuar Garbh Choire (another name for. Garbh Choire Dhaidh) that there was 'just a touch of snow' and later 'very little snow indeed, only one large lump'. Presumably he meant a large patch, for a patch even 5 m long can vanish in a few hours of warm July sun.

In several years, S. Gordon made notes at Garbh Choire Mor just before lasting snowfall (Table 1 and Appendix 1).

His descriptions and photographs (Gordon 1912a, 1925) show that his first, second and third most persistent patches lay in the same hollows as now.

Later he wrote (1963) 'From observations I have made regularly for more than half a century', the snow is the most persistent in Scotland. In fact he did not regularly observe at Garbh Choire Mor for more than half a century, and missed the 1933 disappearance (McCoss, below). In 1920 he went too early to be sure how much survived, though one can make a reasonable inference on this. His detailed size estimates and site coverage in 1910 (Gordon 1912a) were not maintained subsequently, except in six scattered years, when he lived elsewhere and visited the corrie only in 1926, 1937, 1942, 1944, 1945 and 1959. However, he maintained annual contact with Mar stalkers then and since. To study survival until lasting snowfall during a run of years, ideally one must live in the study area during autumn and also observe frequently.

In late August 1918, Bell (1919) saw several patches, the largest covering 250 square yards. On 20 July 1924, Parker (1925) reported many big patches, 'one believed to be permanent'. Alexander (1928) stated that the snow has 'never been known to wholly disappear'.

During a speech at the Cairngorm Club's annual dinner in Aberdeen, Mr McCoss said that in September 1933 'every snowfield in the Cairngorms had disappeared' and 'this is the first occasion in my experience that such a thing has happened' (Notes in Anonymous 1934). The Notes named the corries that he checked, including Garbh Choire Mor. With such a crucial report it is worth probing further. The observer was Club President James McCoss, one of its most active experienced mountaineers (obituary notice, Anonymous 1952). The Garbh Choire Mor snow vanished between 17 September and 1 October, and *The Times* and *Meteorological Journal* reported no snow for the first time in living memory (Cairngorm Club Secretary William A. Ewen, 1946 letter to A. Tewnion). Manley (1971a) wrote that in 1933 it vanished 'for the first time since before 1864'. Gordon (1950) stated that it was reported to him 'about the year 1935' that the snow had gone, and (1951) it is said that it vanished in 1936, but these were errors for 1933.

Of Ciste Mhearad on Cairn Gorm, Forsyth (1894, published in July) wrote 'James Grant, late keeper (Rebhoan), who has known the hill, man and boy, for more than 50 years, says he never saw it without snow; one year there was but a patch left, which he could almost cover with his plaid'. It was known as The Snow House because snow 'is always found' (McConnochie 1896), and on Cairn Gorm's first recorded ascent in 1801, The Snow House was mentioned because 'it is never melted either by the sun or rain', a phrase written by Mrs Sarah Murray (1803) who made the ascent in September. Forsyth (1900, preface written in December 1899) wrote 'Even in the hottest summer it does not altogether disappear', and Cash (1902) that Ciste Mhearad was 'notable as always containing snow'. Alexander (1928, preface January 1928) wrote 'The snow-bed, though in some summers very diminished, is said never to disappear'.

The Abernethy deerstalkers informed S. Gordon (told to AW) that snow always lay there, back to at least 1860. Gordon, whose experience there extended to 1907 (told to AW), never knew it without snow (Gordon 1925), but regarded its survival in 1914 as a close thing. At his request, an Abernethy stalker went there in November 1914, reporting a piece of snow estimated to weigh 7–8 lb. Gordon wrote, 'It may have melted, but the winter snows came to the hills the very next day'. I regard survival as a reasonable inference. McVean's (1963) paper included a photograph of the snow-less hollow in September 1958, and he wrote that snow survived 1962. It vanished in 1969 for the first time since 1959 (Green 1974).

Gordon (1912a) measured surviving snow on Ben Macdui in 1910. Of Ear-choire Sneachdach on Beinn a' Bhuird, he wrote that snow 'as often as not, remains....throughout the year', and survived 1907 and 1909, but not 1908 and 1910. His diaries and notes show survival there in 1914, 1919, and 1920, but not 1921. Of Lochnagar he wrote (1951) of snow lying 'on occasion throughout the year'. This could refer to the west corrie, Coire Lochan nan Eun, where AW has seen it survive.

Hawke (1952) estimated 700 000 and 35 000 square yards of snow in July 1951 and 1952 for Braeriach patches in view from Ben Macdui. Baird (1957) studied the patch north-east of Ben Macdui in 1955 when it just vanished, and 1956 when it survived, in relation to air temperature at a nearby screen. Berry (1967) measured Garbh Choire Mor patches in 1965–66, and stated they 'are reputed to have disappeared' in 1947 (incorrect) and 'the decay of small snow patches can be very rapid towards the end and it is likely that the snow-beds could have passed away unobserved'.

His statement about decay is often correct, but decay tends to slow in October and later, and snow does not vanish unobserved if watchers make effort.

Hudson (1976, 1977 and 1978) measured snow on the central Cairngorms shortly before winter in 1975–77. He wrote of snow that vanished between 16 October 1975 and New Year as most unusual, but AW found that a patch shrank after mid November 1983 and others vanished after mid November 1994 while yet others became noticeably smaller when viewed from the valley during the early few days of December 1994. If the last visit is too early, patches that were thought to have survived may vanish. On 10 October 1977, Hudson (1978) saw patches 51 m and 15.6 m long, and reckoned that both would survive, but AW on 25 October found only one left, < 25 m long.

Spink (1965 and annually till 1980a) noted patches in late July (once August) 1964–78, and published informants' notes which were mostly too early for reliable decisions on survival. His Table 1 in Spink (1980b), entitled 'Summary of snowbed survival (July survey)', lists patches surviving till winter, but July is months too early for decisions about survival. In August 1964 from Ben Macdui he saw 'no signs of snow survivals on Ben Nevis' (Spink 1965), but most of the Ben Nevis patches that last till winter are invisible from Ben Macdui. He stated for 1969 'there appeared to be a clean sweep' (Spink 1970), but later definitely 'the clean sweep of autumn 1969' (Spink 1971). When Pottie (1995) wrote of Garbh Choire Mor snow not melting since 1969, he was referring to Spink's (1971) overstatement. Others correctly observed that snow last vanished there in 1959 (Hudson 1977, Watson *et al.* 1994, Watson 1995).

After visiting in July 1976, Spink (1977) wrote that Garbh Choire had 'a tiny quantity at the end of September…I suggest must have finally disappeared during October'. Later he wrote definitely 'The clean sweep of snow by late September 1976' (1978). Although Pottie (1979) recorded two patches surviving in 1976, Spink (1980a) stated in the same journal that none survived 1969 and 1976, and wrote that snow 'just survived' 1971, though his only note from autumn 1971 was 'I understand that small areas survived'. For Cairn Gorm in 1978 he stated 'I feel that none could have survived' (1979), but this changed to no survival (1980a).

He wrote (1980a) that snow at Ciste Mhearad survived only in 1974 and 1977 during 1972–78, but it did in 1972 and 1978 too (Watson *et al.*, 2002). He stated that none survived on Lochnagar, but it did in 1972 and 1974. Of Lochnagar he wrote 'In a year with normal winds no snow is evident on this mountain by the end of July' (1973). Because he observed from Speyside, he missed patches visible only on Deeside, and AW saw snow on Lochnagar on 1 August in 13 of Spink's 16 years in 1964–79. For 1979, Spink (1980b) wrote 'No snow visible from the Cairngorm area' on 22–27 July, but AW saw 37 patches from Cairn Gorm then, and even on 7 August still 10, including three over 25 m long. Spink later wrote to AW, mentioning poor visibility for distant viewing on his visit. For 1979 he reported survival on Beinn Bhrotain, Beinn a' Bhuird and Ben Avon, based on AW's letter to I. Hudson giving AW's observations on 21 October, but AW found none left on 5 November. Hence it would be best to use Spink's papers only for information on those sites that he visited at close range or in good visibility on the stated dates.

Others' published information for Ben Nevis and nearby hills

Gordon (1920) wrote that a patch below the cliffs of Ben Nevis 'has never been known to disappear entirely'. He stated (1948) that it vanished in 1945 for the first time in living memory, evidently unaware that it had gone in five earlier years. Later (1951) he wrote that there is a good view of the snow-bed in Observatory Gully from the Caledonian Canal, but one cannot see it from the Canal because Tower Ridge obscures it. The most persistent patch visible from the Canal lies at the foot of Point Five Gully, and usually melts sooner than the one in Observatory Gully. However, Gordon correctly noted that the snow in Observatory Gully can be seen from the ridge to the east (i.e. Carn Mor Dearg), and indeed all snow on the north-east cliffs can be seen from that ridge. He wrote that mountaineers there in the first days of November 1945 saw the Ben Nevis cliffs 'entirely free of snow', after the snow-beds had already been observed to be small at the end of October.

Manley (1949) cited a report (*Scottish Mountaineering Club Journal* 10, for 1908–09) that snow at the top of Observatory Gully on Ben Nevis disappeared in late September 1908, but other snow-beds on the hill survived in that year. He omitted this record when stating (1969), 'All the evidence indicates that 1933 was the first summer for a quite indefinite period in which the most persistent of the drifts, that in the Observatory Gully at 3800 feet on Ben Nevis was known to have completely disappeared'.

His account (1971a) rested partly on his visits to Ben Nevis from 1924 onwards, and on a 'Snow Book' that he put in a hut below the cliffs in 1938–70, for notes by climbers. He warned that a small remnant might linger in Gardyloo Gully where it would be seen only by the few climbers tackling that gully. However, observations by several individuals over many years (above) show that snow there invariably vanished earlier than at Observatory Gully. Also, all of Gardyloo Gully is easily seen from Carn Mor Dearg. Manley noted snow-less autumns on Ben Nevis as 1933 (the first since before 1840), 1935, 1938, 1940, 1945, 1949 (virtually certain), 1953, 1958, 1959 and 1969, and 'evidently a very close thing in 1957, 1960 and 1970'.

Dansey (1935) wrote that in September 1933 the Ben Nevis snow-bed had disappeared by the 22nd, whereas a patch under Aonach Beag seems to have held out until about the 30th. The snow-beds became very small in autumn 1934, but James McCoss told Dansey of two small beds still under Braeriach on 20th September and a Fort William correspondent informed him that they did not vanish under Ben Nevis.

Macphee (1936) observed the 1935 disappearance. Crocket (1986) stated that 1935 was notable for 'complete absence of snow on Ben Nevis for only the second time in living memory; or at least within the memory of a deceased local resident who had been born in 1840. On September 28, Macphee visited the places on Nevis where patches of snow were usually to be found, and there were none. The one other occasion when a complete absence of snow was recorded was in 1933'.

Shaded by tall steep crags, the gullies of Ben Nevis gather much snow blown off the plateau or falling in avalanches (Richardson *et al.* 1994). In the cold 1951, snow also survived away from gullies, such as a 225-ft patch that partly covered Lochan Coire na Ciste on 16 September (Hawke and Champion 1952).

Champion (1952) visited on 16–17 June, and Spink (1955 and later references) in late July (August in 1953–54). This is far too early to note survival, except in years when all snow has already gone, as at Observatory Gully in August 1953 (Spink 1955) and autumn 1969 (Gordon, in Spink 1970).

Spink (1980a) gave 1971 as a year of vanishing on Ben Nevis, but his 1971 report (1972) reported a local observer seeing small patches in mid October. They would have survived, for on 12 October came lasting snowfall (AW). Reporting that Manley had seen a big patch in September 1972 from the Fort William-Spean Bridge road, Spink (1973) wrote 'This was probably the usual bed in the Observatory Gully', but it cannot be seen from that road, whereas the Point Five patch is visible. In 1976, all snow had gone by 22 August (Hudson 1977).

Dansey (1905, 1919 and 1920) gave good descriptions of the approximate locations of long-lying snow patches on Ben Nevis, Aonach Mor and Aonach Beag, and his accounts for the last two hills surpass anything since, up to the last decade. Hence it seems surprising that later writers on snow patches, again up to the last decade, apparently ignored Aonach Beag and did not mention the locations on Aonach Mor. Writing of the view from Glen Roy, Gordon (1951) remarked that the great snowfield along the upper slope of Aonach Mor 'is a prominent feature of the landscape until late in summer'. Of Aonach Mor, Manley (1971a) stated that patches survive in many years despite 'less opportunity for drift-accumulation' than Ben Nevis. In fact it has a much bigger smoother plateau than Ben Nevis, and in some autumns the eastern corries hold a bigger area of snow (observations above). Dansey (1919) noticed in September 1918 that the remnants of a fresh snowfall on 31 August lay around the top of Ben Nevis, but less than on Aonach Mor or Aonach Beag. He judged that this is probably because the Aonachs have smooth tops conducive to drifting, whereas Ben Nevis has a top with boulders, and so drifting cannot take place much till the snow has attained some depth.

Previous reviews of Scottish snow survival

Manley wrote (1952) of disappearance in 1933, 1935, 1938 and 1945 'of one or more of the drifts' at Ben Nevis and Braeriach, but 'or more' and 'Braeriach' were incorrect for 1935, 1938 and 1945. In a section on Scotland he wrote (1969) of 'the most persistent of the drifts, that in the Observatory Gully', but snow on Braeriach was far more persistent then, and still is. After mentioning the 1933 disappearances on Ben Nevis and Braeriach, he stated (1970), 'since then there have been at least seven occasions when the last of them melted completely in the late summer'. Though correct for Ben Nevis, for Braeriach this was again an error, broadly repeated since by Lamb (1977) and Lockwood (1982), who cited Manley (1971a) without doing personal fieldwork.

Hence all three authors overstated the frequency of disappearance of all snow. Lamb wrote correctly that all vanished in 1933 'for the first certainly known time', but continued incorrectly except for 1959, 'This occurred in three years each decade from the 1930s to the 1950s', with 'one rather uncertain case in the 1960s'. He gave 1933, 1935, 1938, 1940 (uncertain), 1945, 1949 (probably), 1953, 1958, 1959, and 1969 (almost certain). These comments apply to Ben Nevis, but are incorrect and overstated for Braeriach, where all snow melted in these decades only in 1933 and 1959. Lamb's account differed somewhat on 1940 and 1949 from Manley's (last section), but he did no fieldwork, so it is reasonable to regard Manley's account as more reliable.

Lamb wrote that the positions of surviving snow vary with the depth 'accumulated in winds from this or that direction'. In fact, however, positions vary little from year to year, as shown by detailed evidence by colleagues and myself in the field over many decades. The evidence has included photographs, and also detailed plotting on 1:10 000 maps in the field. These show that each patch's centre did occasionally alter slightly between two years when the patch was similar in size, but when this occurred it was so unusual that I made a special note about it. The main part of the snow patch and the most persistent part lay in the same position over the years, filling an obvious hollow. This allowed us to give for each site a precise map reference, altitude, aspect, and topographic type.

Of the two exceptional locations with persistent snow on Ben Nevis and Braeriach, Manley (1952) stated that the drifts lay 'in the sheltered and deeply shaded gullies descending steeply below the summit ridge', and wrote (1969) of the 'snowbed in the gully facing north-east at the head of the Garbh' (sic) Choire on Braeriach', and (1970) of the snow in steep, narrow and much shaded north-facing gullies (later repeated by Lockwood), and stated (1971b) that the site at Observatory Gully faces north and 'Even at midsummer, direct sunshine is virtually absent'. In fact, the most persistent snow at Observatory Gully faces north-east, and the comment about sunshine is overstated. The most persistent patches in the Garbh Choire (or more accurately the Garbh Choire Mor) lie at a cliff-foot, not in gullies. Also they face north-east, not north, and are open to summer sunshine until the afternoon.

Evidence on trend over the decades

At Ben Nevis, no snow survived 1935, 1938, 1940, 1945, 1949, 1953, 1958, 1969, 1976, 1997–98, 2001, 2004–05 and 2007 when it survived at Garbh Choire Mor, or 1933, 1959, 1996, 2003 and 2006 when it vanished at the latter too. Ben Nevis snow survived all 93 years in 1840–1932, but only 58 of the 77 years 1933–2010, a big fall in proportion (analysing this case and others below using Yates' correction, $\chi^2 = 23.41$, $P < 0.0001$). Allocating a score of 1 for a survival year and 0 for vanishing, I compared the annual score with the year. They were negatively related ($r_s = -0.352$, $n = 170$ years, $P < 0.0001$). Snow disappearance occurred during three years in the 1930s, two in the 40s, three in the 50s, and one in the 60s, but only one in the 70s, none in the 80s, three in the 90s, and five in 2000–10. This suggests more persistence in the 60s–80s than in the 30s–50s plus 90s (Mann-Whitney $U = 0$, $n = 3$ and 4 decades, $P = 0.056$), and more than in 2000–10.

Hodgkiss (1994) wrote that Aonach Mor's corries 'hold snow throughout most years', and 'It is claimed that Roybridge railway station is the only one in Britain from which snow can be seen all year round' (one can see most of the late patches on Aonach Mor and Aonach Beag). The quotation repeats that in earlier editions of the book, and should not be taken to refer to years since 1990. Snow survived on these hills in only 10 years during the 23 years 1987–2010.

In Garbh Choire Mor, snow had not been known to vanish within living memory (Gordon 1909). Up to the 1933 disappearance it had not vanished in living memory since before 1864 (Manley 1971a). In 1944, S. Gordon told AW that Mar deerstalker Charles Robertson, who for many years lived at Corrour Bothy near the snowfield from spring till winter, informed him in 1907 that snow had not gone, back to 1850 at least. Other Mar stalkers (Appendix 2) confirmed this to AW. Burton (1863) reported perpetual snow on the Dee side of Braeriach in 1847 and an unusually snowy 1836, and described hill-walking in the Cairngorms back at least to autumn 1829, so it is reasonable to regard snow persistence as extending at least back to 1829, albeit not explicitly recorded by an observer and author until Gordon.

In 1904–22, Gordon noted numbers and lengths of surviving patches there in four years (Table 1). In several years when he did not give such data, his descriptions and photographs allow one to infer an annual score for the extent of surviving snow. My scores for his 1904–20 observations exceeded scores using the same method on data since 1933 (means 3.91 and 2.53, medians 4 and 2, Mann-Whitney $U = 134$, $n = 11$ and 68 years, $P = 0.0004$). In 1910, a year when he found less snow surviving than usual, the two largest of four patches totalled 75 yards in length (Gordon 1912a) or 69 m. Since 1933, the mean annual number and total length of all patches (in the 73 and 69 years respectively when these were recorded) have been 2.85 and 62.15 m. Hence the later period had even smaller mean values than what Gordon regarded as less than usual up to 1910.

All snow vanished at Garbh Choire Mor in five years in 1933–2010, and the difference in proportion compared with 1829–1932 was significant ($\chi^2 = 4.74$, $P = 0.0295$). However, the correlation of survival with the calendar year was low ($r_s = -0.195$, $n = 181$ years, $P = 0.0087$), and likewise for the years of recorded observation of individual patches since 1904 ($r_s = -0.167$, $n = 106$, $P = 0.088$). Nonetheless, survival scores in 1829–1932 significantly exceeded those in 1933–2010 ($U = 3744$, $n = 104$ and 77, $P = 0.0087$). The annual number of surviving patches was not associated with the year ($r_s = 0.0377$, $n = 73$ years, $P = 0.75$), and likewise with total length ($r_s = 0.067$, $n = 69$, $P = 0.59$). This lack of relationship was due to less snow surviving in the 40s and 50s than in later decades.

At the most persistent snow-bed below Sphinx Ridge, snow survived all 23 years 1904–26. However, survival occurred in only 72 of the 78 years in 1933–2010 (Table 1, and Watson et al. 2002). An analysis of this shows no significant decline in the proportion of years survived ($\chi^2 = 0.76$, $P = 0.38$). Likewise, the correlation with the year was low ($r_s = -0.142$, $n = 107$, $P = 0.14$). The snow vanished in one year of the 1930s, two in the 1950s, none in the 1960s to 1980s, and one in the 1990s.

Gordon (1925) wrote, 'by October, if the season be a mild one, it has split up into three separate snow-beds'. Snow at the second-most persistent bed (below Pinnacles Buttress) survived all 23 years 1904–26, but only 56 of the 70 years with such data in 1933–2010 (Table 1, and Watson et al. 2002), Analysis shows a significant decline in the proportion of years survived ($\chi^2 = 5.21$, $P = 0.0225$). The Pinnacles snow vanished in at least one year in the 1930s, at least two years in the 1940s, and in three years in the 1950s, three in the 1960s, none in the 1970s or 1980s, and two in the 1990s. Its survival showed a low correlation with the year ($r_s = -0.161$, $n = 99$ years, $P = 0.11$), obviously due to better survival in the 1970s–90s than in the 1940s–60s. Its survival was correlated with that of the first-most persistent snow below Sphinx Ridge ($r_s = 0.484$, $n = 101$, $P < 0.0001$).

Snow at the third-most persistent bed at Garbh Choire Mor survived 20 of the 21 years with such data in 1904–26, but only 26 of the 70 years with such data in 1933–2010 ($\chi^2 = 19.55$, $P < 0.0001$). Its survival was negatively correlated with the year ($r_s = -0.288$, $n = 91$ years, $P = 0.0058$). It vanished in at least seven years in the 40s, eight in the 50s, five each in the 60s and 70s, seven in the 80s, four in the 90s, and so far all of the 2000s save 2007, 2008, and 2010. Its survival was correlated with that at Sphinx ($r_s = 0.260$, $n = 93$, $P = 0.012$), and more strongly with that at Pinnacles ($r_s = 0.434$, $P < 0.0001$).

Since 1941, snow has gone at Ben Nevis in 16 years and at Garbh Choire Mor in four of them. In the other 12 years at Ben Nevis it survived at Garbh Choire Mor as one patch in six years, two patches in five years, and three patches in one year. In 1942–2010 inclusive, snow has gone at all other north-east Scottish sites in 30 years, 15 of which coincided with disappearance at Ben Nevis.

Survival at Ben Nevis was correlated with that at Garbh Choire Mor in the same year ($r_s = 0.477$, $n = 171$ years, $P < 0.0001$) and with that at Ciste Mhearad of Cairn Gorm ($r_s = 0.594$, $n = 167$, $P < 0.0001$). Survival at Garbh Choire Mor was correlated with Ciste Mhearad survival ($r_s = 0.303$, $n = 167$, $P < 0.0001$).

The number of patches surviving at Garbh Choire Mor was correlated with survival at Ben Nevis ($r_s = 0.542$, $n = 75$ years, $P < 0.0001$) and Ciste Mhearad ($r_s = 0.579$, $P < 0.0001$). Likewise, the total length surviving at Garbh Choire Mor was correlated with survival at Ben Nevis ($r_s = 0.523$, $n = 70$, $P < 0.0001$) and Ciste Mhearad ($r_s = 0.599$, $n = 70$, $P < 0.0001$).

A tradition was that if snow should vanish in Ciste Mhearad, the Grants will lose Seafield estates (Gordon 1925, and Appendix 2). Snow survived there from pre-1844 to 1932, but in only 20 of the 38 years 1933–70, in 16 of the 30 years 1971–2000 (Watson et al. 2002), and not in 2001–09 (Watson et al. 2002–10) or 2010. The fall in proportion from the recorded years 1907–32 (and Appendix 2) to 1933–2010 was large ($\chi^2 = 21.56$, $P < 0.0001$), let alone including survival up to 1906. Survival was correlated negatively with the year ($r_s = -0.555$, $n = 167$ years, $P < 0.0001$). The top decade for survival since 1930 was the 60s, when snow vanished in only one year.

In some recent years the Ciste Mhearad patch was small even during July, in 1992 being only 20 m long on the 12th and vanishing by 31 July. Forsyth (1900) wrote that in summer the stream cut a tunnel in the snow 'some ten feet high and more than a hundred feet in length' and he walked from one end to the other. Since 1946 the stream has cut a tunnel annually, but less than three feet high at the maximum point. This indicates much deeper snow in Forsyth's time.

Gordon (1912a) stated that two snowfields on Ben Macdui (one at upper Garbh Uisge Mor Corrie, and the other unnamed but by his descriptions clearly at Garbh Uisge Beag) 'rarely, if ever, completely disappear', and Alexander (1928) that the top one 'scarcely ever disappears', and of the Loch Avon side of Ben Macdui more generally he wrote of 'great stony wastes from the hollows of which the snow scarcely ever disappears'. In 1942–70, however, the top one survived in only six of the 29 years (AW unpublished), in only six of the 30 years in 1971–2000 (Watson *et al.* 2002), and not in 2001–09 (Watson *et al.* 2002–10) or 2010. Snow at Garbh Uisge Beag survived in only 10 of the 29 years 1942–70 and in a similar 13 out of 40 years in 1971–2010. At both locations this indicates less persistence than in Gordon's and Alexander's time. Of the Feith Buidhe slabs on Ben Macdui, Gordon (1912a) wrote, 'Here are several fields of snow which are usually found....even after the hottest summer'. In more recent decades, snow survived there in only six of the 29 years 1942–70, and only six of the 30 years 1971–2000 (Watson *et al.* 2002), and not in 2001–09 (Watson *et al.* 2002–10) or 2010.

Although Gordon (1912a) wrote that less snow than usual survived 1910, a drift lasted till winter beside a lochan (site 43 in Table 1), where AW saw surviving snow only in the very snowy 1951 and 1967. Also in 1910, surviving patches north-east of Ben Macdui included the one at upper Garbh Uisge Mor, two at lower Garbh Uisge Mor, two at Garbh Uisge Beag, five at Feith Buidhe slabs, and one at the lochan. Since 1971, snow has survived at all the first four locations in the same year only in the snowy years 1986 and 1994 (Watson *et al.* 2002), and since 1942 at all five locations in the same year only in the very snowy 1951 and 1967.

Snow survived in Ear-choire Sneachdach of Beinn a' Bhuird in five of the eight years of Gordon's (1912a, 1925, diary) notes in 1907–21, and 'frequently remained unmelted all through the summer and autumn'. However, AW found it surviving in only three of the 33 years 1938–70, in three of the 30 years 1971–2000 (Watson *et al.* 2002), and not in 2001–09 (Watson *et al.* 2002–10) or 2010. The difference in proportion from Gordon's time is significant ($\chi^2 = 13.52, P = 0.0002$).

Local folk call the snow in Ear-choire Sneachdach the Laird's Tablecloth, and according to legend the Farquharson family will lose Invercauld if it vanishes. The most recent laird Alwyne A.C. Farquharson told AW during a radio interview in mid July 1982 that he knew the story. Here is a verbatim transcript of the BBC tap: AW to Alwyne Farquharson, "Invercauld has a very long history. The original Farquhar got his land from Robert the Bruce didn't he?" Then Alwyne responded: "Yes, I've heard that said. And I've heard it said that so long as the Tablecloth, which is the snow on the top of Beinn a' Bhuird which is the highest mountain on Invercauld, remains spread, the Farquharsons would hold their lands. And that's the reason why no Farquharson will ever admit that the Tablecloth is not spread. They might go so far as to say that the linen was dirty and needed a wash, but if you take the trouble to climb up and have a look you'll find it spread, summer just the same as winter!" Both then laughed.

Gordon (1925) wrote, 'snow is almost always to be found' in Coire Creagach of Monadh Mor. It survived in only two of 33 years in 1938–70, and two of 38 years in 1971–2009 (Watson *et al.* 2002–10) and not in 2010. In 1920, two patches lasted till winter (Gordon 1921), but in 1938–2008 this occurred only in 1951 and 1967.

I compared snow at the above four sites (Ciste Mhearad of Cairn Gorm, the top of Garbh Uisge Mor, Ear-choire Sneachdach, and Coire Creagach) in the 29 years 1942–70 with the 40 years 1971–2010. Snow survived in a total of 26 cases out of a total of 116 (29 x 4) in the first period, and 27 out of 156 in the second. Statistical analysis showed no significant difference in proportion between them ($P = 0.71$). Hence the fall in survival during the 1930s did not continue to yet lower levels in later decades.

Of the other Ciste Mhearad, on Carn Ban Mor of Glen Feshie, MacBain (1890) who was brought up in the glen wrote, 'Here snow may remain all the year round'. Gordon (1925) stated, after 'almost every winter....it lies until late in summer' and (1948) 'the great snow-field....each year lies....until late summer'. Alexander (1928) wrote that it 'often carries snow until late in the season'. In recent decades it vanished by the following dates (not noted in most years), 17 July 1945 (Perry 1948), 3 July 1948, 13 August 1970, 5 July 1973, 28 June 1981, 7 September 1984 and 6 October 1994 (AW). In 1995–96 it had gone by 20 August and 15 August (maybe up to five days sooner, D. Duncan), and in

1997–2006 by 1 July, 5 June, 1 July, 22 June, 8 August, 7 July, 1 June, 6 June, 6 May, 4 July (Watson *et al.* 1998–2007) and 10 June 2007, 6 July 2008, 24 June 2009 (D. Duncan).

Gordon (1925) noted of Horseman's Corrie (grid reference 936 976) on Braeriach that there was 'frequently snow until September', though he did not record it surviving till winter. In 1971–2010 it lasted till September only in 1983, and had gone by 1 August in 20 of the 39 years.

In Coire an Lochain of Cairn Gorm, 'snow-beds linger as often as not throughout the year' (Gordon 1915). In the 30 years 1971–2000, however, snow survived there in only six years, and a second patch in only three of them (Watson, Davison & Pottie 2002). Also, no snow survived there in 2001–09 (Watson *et al.* 2002–10) or 2010.

If one considers survival at all locations in north-east Scotland except Garbh Choire Mor and allocates a score of 1 for survival of any of them, and 0 for survival of none, survival was not correlated with the calendar year ($r_s = -0.032$, $n = 68$, $P = 0.80$). This occurred because survival was at first low, associated with warmer decades in the 1940s and 1950s, and then increased with the coming of colder decades from 1960 to the mid 1980s. Since 1960 it showed a significant negative correlation with the year ($n = 50$ years, $r_s = -0.488$, $P = 0.0003$). The correlation has been negative again since 1986 ($n = 25$, $r_s = -0.477$, $P = 0.017$).

At all locations in north-east Scotland other than Garbh Choire Mor, the number and the total length of surviving patches have been poorly associated with the year ($n = 55$, $r_s = -0.18$ and -0.270, $P = 0.17$ and 0.21). Since 1986, however, the number and length of surviving patches have been more strongly correlated, negatively, with the year ($n = 25$, $r_s = -0.470$ and -0.544, $P = 0.012$ and 0.0097).

In short, the evidence indicates less persistence since 1930 than in earlier decades and a marked decline since the late 1980s.

Reasons for more persistent snow at Garbh Choire Mor than at Ben Nevis

Because Ben Nevis in winter and spring receives heavier precipitation and more fog than the Cairngorms, snowfall and rime at the top of Observatory Gully exceed those at the snow-beds below the cliffs of Garbh Choire Mor. The 19-year record from the observatory on Ben Nevis summit showed a mean annual precipitation of 410 cm (Thomson 1933), whereas data from the high Cairngorms indicate 230 cm (Green 1974). This difference holds in winter and at all other seasons. Precipitation in winter at both areas exceeds that in summer.

During winter at or near sea level in Scotland, isotherms tend to run north-south, with warmer temperatures in the west near the relatively warm Atlantic, than in the east near the colder North Sea. Also, the centre of Scotland including the Cairngorms has a more continental climate than western Scotland including the Ben Nevis area, with colder winter temperatures (Manley 1952). Manley (1940) reported that the number of days with sleet or snow observed to fall on lowland north-east Scotland (his map showed a big lobe extending from the coast south-west into the Cairngorms) greatly exceeded that in the west Highlands.

Likewise the number of days with snow lying in a median winter during 1941–70, reduced to sea-level, was markedly greater in north-east Scotland, including the Cairngorms, than in the Ben Nevis area (Jackson 1978). Jackson found that the number of days with snow lying at a given altitude in the central Highlands exceeded that in the west Highlands. Days with snow lying at a given altitude tend to be fewer on Ben Nevis than on the Cairngorms (Meteorological Office 1953–92, Manley 1949, 1971a). Hence thaws at a given altitude are more frequent on Ben Nevis, as climbers well recognise. This tends to be counteracted by heavier winter precipitation, including snowfall, than in the Cairngorms.

Both places receive extra snow from avalanches, especially from collapsing cornices, but the ground at both is so steep that most snow in avalanches continues moving to lower altitudes. Moran (1992) suggested that snow at Observatory Gully comes mainly from avalanches, and so he considered it not a real snow-bed like Garbh Choire Mor. However, there is no objective material difference. Apart from the two bowl-like hollows at Sphinx and Pinnacles in Garbh Choire Mor, the general slope below the cliffs of Garbh Choire Mor is steeper than the low-gradient shelf at the top of Observatory Gully, and thus less likely to retain avalanche snow. The main concentration of avalanche snow at Garbh Choire Mor comes to rest on a flat area in the bottom of the corrie at 1040 m altitude, 220 m horizontally down from the longest-lying snow and 100 m in altitude below it.

Crocket's (1986) guide to Ben Nevis includes frequent references to vast quantities of spindrift snow coming down the gullies after being blown off the plateau even in dry weather, to an extent that can make climbing impossible, and this applies to Garbh Choire Mor also. In short, the supply of snow at both sites depends mainly on snow that has been drifted from the plateau above and then falls down the cliff.

Watson *et al.* (2002) found that three topographic factors (fetch, leeward slope and topographic rise, defined and measured as in that paper) accounted for much of the variation in survival of long-lying patches amongst different locations in the north-east Highlands, including Garbh Choire Mor. Using the same methods, I measured each factor in 16 compass directions for the site at Observatory Gully. Totals for fetch, leeward slope and topographic rise were 1630, 2630 and 16 570 m. Values at Garbh Choire Mor had been measured as 4900, 18 950 and 10 280 for Watson *et al.* (2002).

The much larger values for fetch and leeward slope at Garbh Choire Mor than at Observatory Gully are associated with the nearby plateau being much bigger in area and also broader in shape than that behind Observatory Gully. Areas of plateau or gentle slope < 10% gradient behind Garbh Choire Mor amount to 266 ha, excluding the less relevant 30 ha on the dome of Braeriach summit. In contrast, such areas at Ben Nevis occupy only 12 ha, and even including the less relevant Carn Dearg only 45 ha.

The higher value for topographic rise at Observatory Gully than at Garbh Choire Mor is associated with Ben Nevis rising from sea level and being the UK's highest hill. A big rise from sea-level is likely to induce air turbulence and thus increased precipitation. At Ben Nevis, therefore, topographic rise and heavier precipitation would be conducive to more snow at Observatory Gully than Garbh Choire Mor.

The cliff-foot behind the Sphinx and Pinnacles snow at Garbh Choire Mor is steeper than at Observatory Gully, especially that behind the Sphinx patch. Consequently the Sphinx snow sits close to the wall behind, with less room for air behind. The Observatory Gully snow lies on sharp-edged boulders with air circulating among them, whereas the Sphinx and Pinnacles patches lie on subsoil, without air circulating freely below. Unlike the top of Observatory Gully, the Sphinx and Pinnacles patches lie in deep bowl-like hollows, which reduce air circulation and thus the melting rate.

Garbh Choire Mor is a large bowl-shaped corrie, which includes in its far west corner a small inner semi-circular high-altitude corrie, closely tucked into the cliffs and plateau behind. As a result, snow drifted off the plateau and falling down the cliff on to the two longest-lying snow-beds comes from a greater number and variety of directions, in compass degrees from 160 veering via 360 to 14, or about 210 in total. By contrast, snow drifted off Ben Nevis plateau on to the site at Observatory Gully comes from 98 veering to 260, or only about 160 in all.

In addition, rainfall in summer and autumn is much heavier on Ben Nevis and nearby hills than on the Cairngorms. Water is a better conductor of heat than air, and rain percolates down through a snow patch due to gravity, so one would expect a faster rate of melting. The sole exception would be when the upper surface of the snow is so icy that it quickly sheds rainwater instead of the rainwater managing to percolate down into the snow. Moreover, greater wind-speeds in the oceanic climate of the western hills than in the more continental east would add to faster melting.

Discussion

Human-induced and other biological factors that reduce snow patches

Snow fences were erected in 1970–73 to increase snow-lie on ski tows and runs at upper Coire na Ciste on Cairn Gorm. The persistence of surviving snow declined at site 85 in Table 1 of Chapter 5, associated with fences intercepting snow blown on winds from south-west clockwise to north-north-west. Two patches lasted there till mid September 1946, and snow survived 1951 and 1967. Since 1970, however, snow has gone by 1 September in all years and by 1 August in almost all. Possibly these fences may have reduced snow at Ciste Mhearad on winds from west to north-west, though to a lesser extent because its snow lies much further from the fences (the nearest is 400 m).

Since the late 1960s on Cairn Gorm and Ben Macdui, parties from outdoor centres and the armed forces dug snow holes for training in winter survival. Because these increase the area of surface exposed to air, melting must be faster. Until the mid 90s, parties urinated and defaecated on the snow. They used Ciste Mhearad most, Coire Domhain often, and Garbh Uisge Beag or upper Feith Buidhe rarely. I infer that this would not change snow disappearance by more than a week at the most heavily used site, but such a difference might determine whether a patch survived or not in a year of little snow. Another factor in recent decades is more 'black snow' due to distant pollution, which must increase melting slightly.

Snow decline and climatic change

More frequent vanishing of snow since 1930 than beforehand has been attributed to a warmer climate (Manley 1971b, Lamb 1977, Parry 1978, Watson & Cameron 2010). In the present paper, vanishing at Ben Nevis was less frequent per decade in the 1960s, 70s and 80s than in the 1930s–50s and also less than in the 1990s and 2000s. This tallies with colder winters in the 1960s–80s.

Acknowledgements

I am pleased to acknowledge Seton Gordon, Alex Tewnion and my father Adam Watson for photographs and observations, and for observations Patrick D. Baird, David Duncan, Justin Irvine, Malcolm Smith and those in Appendix 2. Thanks are due to John Duff for searching the Cairngorm Club Journal, Dr Richard Davison for commenting on a preliminary draft, Dr Julian C Mayes for reading the manuscript and giving valuable advice, and Dr Brian J. Coppins for identifying lichens (see Appendix 3).

References

Alexander, H. (1928). The Cairngorms. Scottish Mountaineering Club, Edinburgh
Anonymous (1934). Proceedings of the Club. Cairngorm Club Journal 13, 253. Also Notes, 277.
Anonymous (1952). In memoriam James McCoss. Cairngorm Club Journal 16, 273–274.
Baird, P.D. (1957). Weather and snow on Ben Macdhui. Cairngorm Club Journal 17, 147–149.
Bell, J.H. (1919). Notes. Cairngorm Club Journal 9, No 52.
Berry, W.G. (1967). Salute to Garbh Choire Mor. Scottish Mountaineering Club Journal 28, 273–274.
Burton, J.H. (1863). The Cairngorm mountains. Blackwood, Edinburgh.
Cash, C.G. (1902). Days and nights on the Cairngorms. Cairngorm Club Journal 3, 317–326.
Champion, D.L. (1952). Summer snows around Ben Nevis. Weather 7, 180–184.
Dansey, R.P. (1905). Glacial snows of Ben Nevis. Symons's Meteorological Magazine 40.
Dansey, R.P. (1919). A tramp between Lochaber and the Cairngorms. Scottish Mountaineering Club Journal 15, 196–206.
Dansey, R.P. (1920). The Aonachs. Scottish Mountaineering Club Journal 16, 192–193.
Dansey, R.P. (1935). Disappearance of the Scottish snow-fields. Scottish Mountaineering Club Journal 20, 366–367.
Davison, R.W. (1987). The supply of snow in the Eastern Highlands of Scotland: 1954-5 to 1983-4. Weather 42, 42–50.
Forsyth, W. (1894). Outlying nooks on Cairngorm. Cairngorm Club Journal 1, 134–137.
Forsyth, W. (1900. In the shadow of Cairngorm. Northern Counties Publishing, Inverness, reprinted (1999) by Bothan Publications, Lynwilg, Aviemore.
Fyffe, A. & Nisbet, A. (1995). The Cairngorms rock and ice climbs. Scottish Mountaineering Trust, Edinburgh.
Gordon, S. (1908). Braeriach in September. Cairngorm Club Journal 5, No 30 (signed SG, January number).
Gordon, S. (1912a). The charm of the hills. Cassell, London.
Gordon, S. (1912b). Snowfield on Braeriach. Cairngorm Club Journal 7, 179 (signed SG).
Gordon, S. (1915). Hill birds of Scotland. Arnold, London.
Gordon, S. (1920). Land of the hills and the glens. Cassell, London.
Gordon, S. (1921). Wanderings of a naturalist. Cassell, London.
Gordon, S. (1925). The Cairngorm hills of Scotland. Cassell, London.
Gordon, S. (1927). Days with the golden eagle. Williams & Norgate, London.
Gordon, S. (1944). A Highland year. Eyre & Spottiswoode, London.

Gordon, S. (1948). Highways and byways in the Central Highlands. Macmillan, London.
Gordon, S. (1950). Snow flora of the Scottish hills. Nature 165, 132–134.
Gordon, S. (1951). Highlands of Scotland. Hale, London.
Gordon, S. (1963). Highland days. Cassell, London.
Gordon, S. Many years. Diary, held in the Seton Gordon archive at the National Library of Scotland, Edinburgh.
Green, F.H.W. (1974). Climate and weather. In: The Cairngorms (by D. Nethersole-Thompson & A. Watson), 228–236. Collins, London.
Hawke, E.L. (1952). Midsummer snow on Braeriach. Weather 7, 121.
Hawke, E.L. & Champion, D.L. (1952). Report on the snow survey of Great Britain for the season 1950–51. Journal of Glaciology 2, 15–37.
Hodgkiss, P. (1994). The Central Highlands. District Guide. Scottish Mountaineering Club.
Hudson, I.C. (1976). Cairngorm snow-field report 1975. Journal of Meteorology 1, 284–286.
Hudson, I.C. (1977). Cairngorm snowfield report 1976. Journal of Meteorology 2, 163–166.
Hudson, I.C. (1978). Cairngorm snowfield report 1977. Journal of Meteorology 3, 306–310.
Humble, B.H. (1946). On Scottish hills. Chapman & Hall, London.
Jackson, M.C. (1978). Snow cover in Great Britain. Weather 33, 295–309.
Johnstone, H. (1905). The Club on Braeriach. Cairngorm Club Journal 4, 350–359.
Lamb, H.H. (1977). Climate: present, past and future. Vol. 2. Climatic history and the future. Methuen, London.
Lockwood, J.G. (1982). Snow and ice balance in Britain at the present time, and during the last glacial maximum and late glacial periods. Journal of Climatology 2, 209–231.
MacBain, A. (1890). Badenoch: its history, clans and place names. Transactions of the Gaelic Society of Inverness 16, 148–197.
McConnochie, A.I. (1895). The Cairngorm Mountains. The Central Cairngorms. II The Cairngorm Division. Cairngorm Club Journal 1, 366–384.
Macphee, G.G. (1936). Ben Nevis guide. Scottish Mountaineering Club, Edinburgh.
McVean, D. (1963). Snow cover in the Cairngorms 1961–62. Weather 18, 339–342.
Manley, G. (1940). Snowfall in Britain. Meteorological Magazine 75, 41–48,
Manley, G. (1949). The snowline in Britain. Geografisker Ånnaler 31, 179–193.
Manley, G. (1952). Climate and the British scene. Collins, London.
Manley, G. (1969). Snowfall in Britain over the past 300 years. Weather 24, 428–437.
Manley, G. (1970). The climate of the British Isles. Climate of northern and western Europe (ed by CC. Wallen), 81. Elsevier, Amsterdam (cited in Lockwood).
Manley, G. (1971a). The mountain snows of Britain. Weather 26, 192–200.
Manley, G. (1971b). Scotland's semi-permanent snows. Weather 26, 458–471.
Meteorological Office (1953 and annually till 1992). Snow Survey of Great Britain. Meteorological Office, Bracknell.
Moran, M. (1992). Scotland's winter mountains. David & Charles, London.
Murray, S. (1803). A companion and useful guide to the beauties of Scotland, etc (Vol. 2, third edition). J Bulmer & Son, London
Parker, J.A. (1925). Notes. Cairngorm Club Journal 11, 143.
Parry, M.L. (1978). Climatic change, agriculture and settlement. Studies in historical geography. Dawson, Folkestone.
Perry, R. (1948). In the high Grampians. Lindsay Drummond, London.
Pottie, J.M. (1979). Scottish snowbeds in 1976. Weather 34, 81.
Pottie, J.M. (1995). Scottish snowbeds: records from three sites. Weather 50, 124–129.
Richardson, S., Walker, A. & Clothier, R. (1994). Ben Nevis rock and ice climbs. Climbers' Guide, Scottish Mountaineering Club.
Richardson, W.E. (1954). Dirt polygons. Weather 9, 117–121.

Smith, M. (1961). Climbers' guide to the Cairngorms area. Vol. 1. Scottish Mountaineering Club, Edinburgh.

Spink, P.C. (1955). Summer snows around Ben Nevis. Weather 10, 269–271.

Spink, P.C. (1962). Summer in the Cairngorms. Weather 17, 408–409.

Spink, P.C. (1966). Scottish snowbeds in summer 1965. Weather 21, 127–129.

Spink, P.C. (1967). Scottish snowbeds in summer 1966. Weather 22, 298–299.

Spink, P.C. (1968). Scottish snowbeds in summer 1967. Weather 23, 209–211.

Spink, P.C. (1969). Scottish snowbeds in summer 1968. Weather 24, 115–117.

Spink, P.C. (1970). Scottish snowbeds in summer 1969. Weather 25, 201–202.

Spink, P.C. (1971). Scottish snowbeds in summer 1970. Weather 26, 223–224.

Spink, P.C. (1972). Scottish snowbeds in summer 1971. Weather 27, 35–36.

Spink, P.C. (1973). Scottish snowbeds in summer 1972. Weather 28, 162–164.

Spink, P.C. (1974). Scottish snowbeds in summer 1973. Weather 29, 151–154.

Spink, P.C. (1975). Scottish snow survivals, summer 1974. Weather 30, 235–238.

Spink, P.C. (1976). Scottish snow survivals -- summer 1975. Weather 31, 126–128.

Spink, P.C. (1977). Scottish snow survivals -- summer 1976. Weather 32, 303–304.

Spink, P.C. (1978). Scottish snow beds in summer 1977. Weather 33, 278–279.

Spink, P.C. (1979. Scottish snow beds in summer 1978. Weather 34, 158–160.

Spink, P.C. (1980a). Scottish snow-beds in summer: 1979 survey and some comments on the last 30 years. Weather 35, 256–259.

Spink, P.C. (1980b). A summary of summer snow surveys in Scotland: 1965–1978. Journal of Meteorology 5, 105–111.

Thomson, G. (1933). The meteorology of the Scottish mountains. In: General guide-book (ed. by A.E. Robertson), 18-35. Scottish Mountaineering Club, Edinburgh.

Watson, A. (1995). More data on Scottish snowbeds. Weather 50, 356.

Watson, A. (1998). The Cairngorms. District Guide. Scottish Mountaineering Club.

Watson, A. & Cameron, I. (2010). Cool Britannia. Paragon Publishing, Rothersthorpe.

Watson, A., Cameron, I., Duncan, D. & Pottie, J. (2010). Six Scottish snow patches survive until winter 2009/2010. Weather 65, 196–198.

Watson, A., Cameron, I., Duncan, D. & Pottie, J. (2011). Six Scottish snow patches survive until winter 2010/2011. Weather 66, 223–225.

Watson, A., Davison, R.W. & French, D.D. (1994). Summer snow patches and climate in north-east Scotland, U.K. Arctic and Alpine Research 26, 141–151.

Watson, A., Davison, R.W. & Pottie, J. (2002). Snow patches lasting until winter in north-east Scotland in 1971–2000. Weather 57, 374–385.

Watson, A., Duncan, D., Cameron, I. & Pottie, J. (2008). Nine Scottish snow patches survive until winter 2007/2008. Weather 63, 138–140.

Watson, A., Duncan, D., Cameron, I. & Pottie, J. (2009). Twelve Scottish snow patches survive until winter 2008/2009. Weather 64, 184–186.

Watson, A., Duncan, D. & Pottie, J. (2006). Two Scottish snow patches survive until winter 2005/06. Weather 61, 132–134.

Watson, A., Duncan, D. & Pottie, J. (2007). No Scottish snow survives until winter 2006/07. Weather 62, 71–73.

Watson, A., Pottie, J. & Duncan, D. (1998). Survival of two snow patches in the UK until winter 1997/98. Weather 53, 155–158.

Watson, A., Pottie, J. & Duncan, D. (1999). Only one UK snow patch lasts until winter 1998/99. Weather 54, 369–374.

Watson, A., Pottie, J. & Duncan, D. (2000). Six UK snow patches last until winter 1999/2000. Weather 55, 286–290.

Watson, A., Pottie, J. & Duncan, D. (2001). Forty-one UK snow patches last until winter 2000/01. Weather 56, 404–407.

Watson, A., Pottie, J. & Duncan, D. (2002). Two UK snow patches last until winter 2001/02. Weather 57, 219–222.
Watson, A., Pottie, J. & Duncan, D. (2003). Five UK snow patches last until winter 2002/03. Weather 58, 226–229.
Watson, A., Pottie, J. & Duncan, D. (2004). No Scottish snow patches survive through the summer of 2003. Weather 59, 125–126.
Watson, A., Pottie, J. & Duncan, D. (2005). One Scottish snow patch survives until winter 2004/05. Weather 60, 165–166.
Watson, A., Pottie, J. Rae, S. & Duncan, D. (1997). Melting of all snow patches in the UK by late October 1996. Weather 52, 161

Table 1. Number of snow patches surviving at Garbh Choire Mor, and their total length.

Year	Number	Length (m)	Year	Number	Length (m)
1908	4	161	1954	1	6
1909	3	90	1955	2	25
1910	4	90	1956	4	45
1933	0	0	1957	2	29
1934	2	-	1958	2	36
1937	3	-	1959	0	0
1942	2	-	1960	1	8
1943	2	-	1961	2	26
1944	2	-	1962	3	65
1945	1	6	1963	3	51
1946	2	15	1964	2	17
1947	4	60	1965	4	95
1948	2	23	1966	1	14
1949	1	3	1967	13	400
1950	1	10	1968	3	50
1951	22	420	1969	1	8
1952	2	20	1970	3	50
1953	1	9			

For 1933 see text and Anonymous (1934), for 1934 see Dansey (1935), for other years in 1908–42 and 1945 see S. Gordon (Appendix 1), for 1943–70 there are notes by AW.

Appendix 1

Gordon's early notes at Garbh Choire Mor compared with since 1942

His first-hand experience there extended to 1904 (Gordon 1963), but he was in close touch with long-resident local deerstalkers. On several occasions (1912a, 1912b, 1925) he stated that the snow has never within living memory been known to disappear.

He (1908) wrote of much snow in 1907. On 1 September fresh snow lay, and old snow on either side of the top of Dee waterfall (AW saw old snow there as late as this only in 1951 and 1967). To judge from his notes on Beinn a' Bhuird (Gordon 1912a), 1908 was almost as snowy. A photograph (Gordon 1921) taken in 1908 (told to AW) showed four patches that AW measured against the known height of the cliff, estimating a total of 160 m. Four patches survived 1909, when Gordon (1912a) wrote that snow 'had dwindled more than….for a considerable number of years'. I infer that 1904–08 had considerably more than 1909, and the 'considerable number of years' (obviously from observations told to him by local deerstalkers) puts this back further than 1904. He measured (1912a) surviving 'snowfields' in 1910, the two main ones 45 and 30 yards long, and 'on either side two small fields', so I estimate a total of 100 m. The autumns of 1910 and 1914 had less than usual, with 1910 'wonderfully similar' to 1909, but patches in 1911, 1912 and 1913 were bigger (Gordon 1912a, b, and diary), and 1914 similar to 1910. On 11 October 1921 the third most persistent patch had gone, which he 'had never before seen' (Gordon 1925).

In autumn 1926 the snow in Garbh Choire Mor was the least he had seen there (Gordon 1927). On 23 October 1937 he found three patches (Gordon 1948) and on 22 October 1942 there were two, smaller than he had ever seen (Gordon 1944). In 1945 he found only one patch, the first time he had seen just one.

The number and size of patches in 1909–10 and 1914 exceeded those since 1942. Yet he wrote of less snow than usual in these three years. This suggests that his years of less snow had more and bigger patches than in autumns since 1942, with less snow than usual for this later period.

I scored surviving snow as 0 to 6 (0 for none and 1 for one patch, neither of which he saw in 1904–22). This gave 2 for 1921, 3 for 1909–10, 1914, and 1937, 4 for 1907–08 and 1911–13, 5 for 1920, and 4 for 1904–06 because snow in 1909 had 'dwindled more than….for a considerable number of years'. The scores for 1904–14 and the 1920s averaged the same, about 3.7.

Appendix 2

Information from long-resident local people

Since the late 1700s, the main season for stag shooting on Mar and Abernethy deer forests has been August to late October, with hind shooting in November-January. Local stalkers were on the hills almost daily in the deer season. Mar stalkers told AW in 1942–47 about snow survival at Garbh Choire Mor in earlier years as seen by them. They were Charles Grant and his son Willie, Ian Grant, Sandy and Donald McDonald and their hill-guide sister Nell, and Frank, Robert, Ronald and Walter Scott. Their experience extended to 1880 in Charles' case. Also they gave information passed down by the fathers of stalkers and gillies, and their fathers' colleagues.

Donald Scott, father of the four above sons, died in 1905 at Linn of Dee cottage after 30 years' service with the Duke of Fife, first as a spring-autumn 'watcher' at Corrour Bothy near the snowfield, then at Derry Lodge/Luibeg, and lastly at Linn of Dee, and his experience extended to 1875. A later Corrour watcher was Charles Robertson, whose experience extended to 1848 and died in 1933 aged 99. Gordon's (1920) chapter 11 was largely about Robertson. When Robertson retired, John McIntosh followed him at Corrour Bothy, and Frank Scott in the early 1920s was the last Corrour watcher. Hugh D. Welsh, for many years the Cairngorm Club's president, who was born in 1886 and died in January 1969 aged 82, told AW in 1947 that he often visited Robertson in 1905–12 and McIntosh at Corrour, and saw Garbh Choire Mor each autumn. The above three in the McDonald family moved from Bynack Lodge to Luibeg when their father Sandy senior was appointed stalker there after Donald Fraser died, and the Luibeg beat includes Garbh Choire Mor and many other places from which it is visible.

The Grants said that it had been handed down by generations of stalkers that snow always lay there back to the Earl of Fife's time in the late 1700s (the Earl wrote of a stalker Grant in his daily journal of 1785–92, now in the Special Collections, University of Aberdeen library). The comment that snow had always lain back to the 1700s was less reliable, because it involved unnamed people at second hand.

Pat McLean, born in Abernethy in 1917 and raised there, told AW in 2002 that the legend of the Grants losing Seafield estate if snow vanished at Ciste Mhearad of Cairn Gorm (Gordon 1925) was widely known by local folk in his youth. He confirmed survival there in 1926–32, and recalled his father and colleagues saying snow always lay there. His father was stalker on Seafield's Abernethy which included Ciste Mhearad, and his grandfather stalker on Seafield's Glen More whose march overlooked Ciste Mhearad. Their experience of snow there extended to 1875.

Carrie and Desmond Nethersole-Thompson of Whitewell told AW in 1946 when the snow at Ciste Mhearad vanished in 1926–42 (1933, 1935, 1938, 1940 and 1942). This was confirmed to AW in 1940–42 by former Seafield gamekeeper Robert Grant of Lower Craggan near Grantown, in 1942–47 by Mar men (above) who saw Ciste Mhearad while stalking on Beinn a' Bhuird in September-mid October, in 1946 by Abernethy forester, naturalist and writer Willie Marshall, in 1980 by Dolina Macdonald (wife of the stalker who lived at Glenmore Lodge in 1917–47 and whose beat extended almost to Ciste Mhearad), and in 1993 by Donnie Smith of Lurg who gathered sheep each autumn on land including Ciste Mhearad and also on A' Choinneach looking across to Ciste Mhearad. At the age of three, Smith came in 1917 to Guislich farm below Cairn Gorm, where his father had a grazing tenancy on Braeriach in the west Cairngorms. They moved to Lurg in 1936.

Appendix 3.

Descriptions at Garbh Choire Mor when most snow had vanished

On 11 October 1921, the third most persistent patch had gone (Gordon 1925). Stones on the bottom of the snow-less third hollow were black with dirt, and moss on the bottom was already green with fresh growth. The two most persistent snow patches had been almost black with dirt on 25 September 1921, the Sphinx patch more earth-stained than the Pinnacles one.

On 13 September 1959, when all snow had gone, my father Adam Watson senior described the place. Dry dirt covered boulders and soil in the bowls. It gave the rough boulders a smooth appearance of blue-grey hue. Apart from the dirt layer, soil in and just outside the bowls was subsoil of granite granules and smaller particles. The snow-less bowls at Sphinx and Pinnacles held no plants or insects, and moss formed the sparse vegetation outside the bowls. The cliff foot and the boulders and subsoil just outside the bowls had the pink-orange hue of virgin Cairngorms granite. This was because they were devoid of rock lichens, as noticed earlier by Gordon (1944, 1951).

Alex Tewnion visited the snow-free bowls on 7 October 1959. Scraping through black dirt on the bottom of them, he found no plants below, and none on the surface, except for bits of loose moss that had fallen from above. At the rear of the snowless scoop at Sphinx, black dirt lay an inch thick. Below that, digging revealed sand with small stones, next clay an inch thick, and then stones. David Duncan in 2003 estimated the extent of ground with no lichen or moss as 29 x 14 m at the Pinnacles hollow and 12 x 10 m at the Sphinx one, and the bare stones were loose and unconsolidated.

On 27 September 1961, Tewnion noted that the snow was dirty with earth and debris that had fallen from above. On 26 September 1964 he measured the Pinnacles snow as 35 x 9 x 2 ft and Sphinx 24 x 21 x 4 ft. Moss grew on grit up to 6 ft from the Pinnacles patch and 3 ft from the Sphinx one. This was a few feet beyond the outer edge of the dirt-covered soil and boulders. Where his fingers touched the snow, they were 'black as though with soot, and the dirt in the snow-ice is evidently very fine black soil'. Dirt is common on snow patches (Wilson 1958).

As on rocks beside a glacier, the bottom few metres of bedrock at the foot of Sphinx Ridge and Pinnacles Buttress carry no rock lichens (AW). Lichen-free cliff on Sphinx (estimated by David Duncan in 2003 as extending 20 ft up the slab) extends higher than on Pinnacles, and hardly occurs above the 3rd, 4th and 5th most persistent patches. Other cliffs in the Cairngorms lack this lichen-free band.

Apart from the lichen-free band at the foot, the Garbh Choire Mor cliffs are unusually green from rock lichens. The rock 'bears a patina of lichen which apparently thrives on granite covered for long periods by snow and is responsible for the remarkable greenish hue of the buttresses which is most pronounced when they are seen in contrast with the snowfields and gullies' (Smith 1961). The green hue is strongest on the lower third of the cliff at and near Sphinx and Pinnacles, and also on the lower half at the She-Devil's Buttress, but occurs to a lesser extent up to the topmost cliffs in the corrie. It occurs on other Cairngorms crags where snow lies late, but only on a band on the bottom part, such as in Coire Sputan Dearg. In 2002, David Duncan took rock specimens from Garbh Choire Mor so that lichens could be identified. He collected a piece 17 x 10 x 8 cm, and Dr Brian J. Coppins (Royal Botanic Garden, Edinburgh) identified the lichens. Commonest was the yellowish-green map-lichen *Rhizocarpon geographicum*, others being *R. lecamonnum, R. cinereonigrum* (a species which elsewhere is mainly found in areas of late snow-lie), *R. polycarpum, Lecanora polytropa, Mirquidia complanata, Porpidia crustulata* and *Umbilicaria cylindrica*. On 8 November 2004, Keith Miller (note to AW) noted that the zone of abundant *Rhizocarpon geographicum* above the Sphinx snow-patch began about 15 m up the cliff, possibly 20 m up.

Grit and boulders at surviving patches that are less persistent (e.g. Garbh Uisge Beag and Ciste Mhearad) also lack rock lichens and have the pink colour of virgin Cairngorms granite. However, they support tiny lumps of moss, unlike the vegetation-free grit and boulders at Sphinx and Pinnacles. It is of the *Polytrichum sexangulare-Kiaeria starkei* snow-bed community U11 in Rodwell (1992). It occurs on flat and gently-sloping tops of boulders, but not on steeply-sloping or vertical parts, which explains why it does not grow much on steep bedrock on cliffs.

In October 1959, dung of ptarmigan (*Lagopus mutus*) abounded on the scoop that had held the Sphinx snow (Alex Tewnion, letter to AW). On recent October visits, David Duncan has noted unusually much ptarmigan dung on the ground beside the snow. In October, when other corries were snow-free, Gordon (1925) noticed that ptarmigan made 'fresh roosting hollows....on the icy surface'. AW often saw fresh dung piles there, each from a ptarmigan roosting on the snow. As the snow melts, each pile becomes dispersed as a set of individual faeces over a bigger area of snow, ending as scattered separate droppings on the ground after all snow has gone. In contrast, ptarmigan roosting off the snow leave concentrated piles on the ground, which do not become dispersed. The amount of faeces per unit area is far greater beside the long-lying snow than elsewhere.

References for Appendix 3

Gordon, S. (1944). A Highland year. Eyre & Spottiswoode, London.

Gordon, S. (1951). Highlands of Scotland. Robert Hale, London.

Rodwell, J.S. (Ed) (1992). British plant communities. Vol. 3. Grasslands and montane communities. Cambridge University Press.

Wilson, J.W. (1958). Dirt on snow patches. Journal of Ecology 46, 191–198.

Chapter 5. Snow lasts till winter, north-east Scotland 1942–2010, west Scotland 1945–2010

Summary

Watson *et al.* (2002) listed 55 north-east Scottish snow patches that remained till lasting winter snow in one or more years in 1971–2000, and below I bring this up to 2010. Also I list 96 extra patches there in 1942–70, and 61 in west Scotland during 1945–2010 including two noted by Pottie (1995). For each location I give the map grid reference, topographic type, altitude, aspect, and some data on survival.

Study areas and methods

Watson *et al.* (1994) documented patches in summer-autumn 1974–89 on a large study area covering almost all land east of the A9 Perth-Inverness road, including the Cairngorms massif between Aviemore and Braemar, and the east-west Mounth range south of Braemar. On that area plus the Monadh Liath and Creag Meagaidh hills as seen from the Cairngorms, Watson *et al.* (2002) noted survivals in 1971–2000. Both papers describe Methods. Here I report on this larger area (called north-east Scotland) in 1942–70, and on west Scotland in 1945–2010. For maps showing the main hills, see Watson *et al.* (1994) and Watson *et al.* (2007).

The 1942–70 data should be considered minimal because my total surveys before 1971 were biased towards years when absence of old snow was easily noticed and therefore written by me in notes. During autumns with much old snow (1951 and 1967), I made no total survey, yet noted more survivals than in any year of the 1970–2000 period when I did carry out total surveys. Doubtless I missed some in these earlier snowy years, perhaps many in the west Cairngorms and Creag Meagaidh.

The west-Scottish data are very minimal because of frequent poor visibility to me viewing from the Cairngorms or because I seldom visited sites that cannot be seen from there. So, if a year is not mentioned, this signifies no data, not a lack of snow patches. In 1996 and since, total coverage has been achieved in the west, largely due to a greatly increased number of voluntary observers living locally or visiting.

Unless actually visited, a very small patch can easily be overlooked by the naked eye, especially in haze, rain or fog. A report of no snow should be taken as accurate only if the observer stands beside it; or uses binoculars in good visibility, and, it should be stressed, with detailed knowledge of the exact location.

There is doubt about some past reports on survival, owing to such insufficient effort. Visits in late July such as Spink's (1980) are far too early. Visits even in mid September, such as Manley's (1971a), are too early in years when old snow is still lying. A photograph of the patch at the top of Observatory Gully on Ben Nevis 'at minimum, September 22, 1948, with the first autumn powdering of snow' (Manley 1952) was not in fact at the minimum. Fresh snow did fall lightly on the hills on 21–23 September, but had vanished on 24 September in very mild air. Several days in the next few weeks were summer-like on the high hills, where the first lasting snow came only on 16 October. Even that date is too early in years such as 1983, 1994, 2004 and 2009 when patches disappeared in late October or November, and other patches even at the start of December in 1983 and 1994.

Another drawback of few visits is that an observer may greatly overestimate the length of snow patches thought to have survived. For example, snow patches that survived in 1963 at Garbh Choire Mor had a total length of 51 m (Table 1), as found by me in several visits during September. Then there came snowfalls with much drifting, followed by a thaw and next a frost with a light fresh snowfall. On 14 October the snow from the recent snowfall had turned grey and hard, while the fresh snowfall was white and soft. Had there been no earlier visits, an observer might have inferred a total length of 220 m, based on the grey snow, but it completely covered the 51 m of old snow from the previous winter. In short, there is no substitute for frequent visits throughout autumn and early winter. Indeed, residence in the region, without holidays elsewhere at these seasons, is necessary if there is only one observer. A great advantage since 1996 is that there have been increasing numbers of keen voluntary observers, now in close touch with one another through the internet and also avoiding duplication.

Results

1942–70 survivals at north-east Scottish sites additional to Watson et al. (2002)

Besides 55 patches that lasted till winter in 1971–2000 (Watson *et al.* 2002), snow survived in at least 96 more sites in 1942–70 (Table 1).

In early 1947, much snow fell on south-east winds, and patches that faced north-west survived at Carn an t-Sagairt Mor and Coire Lochan nan Eun west of Lochnagar, March Burn on Ben Macdui, and Coire Cas on Cairn Gorm. The early months of 1940, 1941 and 1942 also had a high proportion of easterly wind, and consequently more snow than usual on the above sites well into July. Snow at Coire Cas lay till 5 September 1972 after another winter with much south-east wind, and two patches on 1 October 1994 had gone by the 7th.

More patches survived 1951 than any year in 1938–2010 inclusive. Nine lay at <900 m, the lowest at 740 m on Ben Avon. The biggest surviving patches in 1938–2010 were in 1967. On Ben Macdui and Cairn Gorm they totalled >1700 m long (see some of them in Eric Langmuir's photographs in Spink 1968). At Alltan na Beinne on Beinn a' Bhuird, men skied on 780 m of snow in late August 1967 and a patch survived. Snow lay there in 1975 till late September, but vanished by 1 October.

Snow survived on the Monadh Liath only at Carn Ban in 1951, and though >100 m long in 1994 on 19 August it vanished in late September. None survived on the Mounth hills from Dalwhinnie eastwards along the Dee's south watershed, save at An Sgarsoch, Lochnagar and Glas Maol. Snow at Brudhach Mor on Glas Maol usually lay into August, in 1983 till mid September, and in 1951 and 1967 till winter. In 1993 it dwindled quickly from 700 m long on 1 July to 35 m on the 17th and nil by the 24th.

1942–70 survivals at north-east Scottish sites in Watson et al. (2002)

Watson *et al.* (2002) noted surviving north-east Scottish patches in 1971–2000. Here I record some of them surviving in 1942–70, and a few back to 1938.

At sites other than Garbh Choire Mor, all snow vanished in 1942–43, 1945–46, 1948–50, 1953–54, 1957–59 and 1969. Autumn 1949 had very little old snow. By 24 August it had gone on Lochnagar, Ben Avon, Beinn a' Bhuird, Beinn Bhrotain, Monadh Mor, Beinn Mheadhoin and Ben Macdui's Coire an Lochain Uaine, by 16 September on other parts of Ben Macdui and Cairn Gorm's Ciste Mhearad, and by the 26th on all of the Spey side of the Cairngorms. Because of few visits in these early years, dates of vanishing may have been considerably earlier.

In 1942–70, snow at Garbh Choire Mor survived each year except 1959. It lasted at Ciste Mhearad on Cairn Gorm except in 1942–43, 1945–46, 1948–50, 1953–54, 1957–59 and 1969–70, and became tiny in 1960, 1961 and 1964. In 1938–70 at Ear-choire Sneachdach on Beinn a' Bhuird it survived in 1951, 1967 and 1970. In 1951 and 1967 it lasted at Coire Creagach on Monadh Mor, Coire an t-Sneachda on Beinn Bhrotain, Coire an Lochain Uaine on Ben Macdui, Coire an Dubh-lochain on Beinn a' Bhuird, Allt an Eas Mhoir on Ben Avon, Coire an Lochain Uaine on Cairn Toul, and all sites noted in Watson *et al.* (2002) at Garbh Choire Mor, at Garbh Uisge Mor, Garbh Uisge Beag and Feith Buidhe on Ben Macdui, and on Beinn Mheadhoin. At Garbh Uisge Beag and low Garbh Uisge Mor it lasted also in 1947, 1956, 1962–63, 1965–68 and 1970, and at high Garbh Uisge Mor also in 1956, 1962–63 and 1970. Snow survived on the slabs at low Feith Buidhe in 1962, 1965 and 1968.

Survival in west Scotland

Table 2 records 61 sites in west Scotland, including the west Highlands and north-west Highlands. Probably more patches survived in the snowy years 1951, 1967 and 1994, when observer coverage there was very incomplete, and maybe in other years. Moreover, snow may have survived on extra hills not in the Table, and on hills that are in the Table but at extra sites on them.

On Na Gruagaichean in the Mamore hills north of Kinlochleven, snow survived in 1951 and several other years

(A. Gorham as told to AW), but not since 1995. In 1993 on Stob Coire Sgreamhach at Glen Coe, I. Cameron found a patch that lasted until the first snowfall in October. It is a reasonable inference that it survived till lasting winter snow. On the Cairngorms, new snow from the first fall on 6 October continued to lie on all 16 of the patches of old snow until the arrival of heavy winter snowfall.

Ben Nevis has long been known as a site of long-lasting snow. Snow at the top of Observatory Gully survived 1946–48, 1950–52, 1954–57, 1960–68, 1970–75 (S. Gordon, as observed by him or told to him by local residents in 1945–76, including letters that he sent to AW or information that he told to AW, and as observed by AW in 1948, 1951, 1954, 1967, 1972 and 1974–75), 1980–95 (D. Scott, I. Sykes, T. Cardwell, S. Fraser, M. Tighe, AW and JP), and in 1999–2000, 2002, and 2007–09 (Watson *et al.* 2000–10) or 2010. On 20 September 1973 shortly before heavy snowfall, J. Duff (told to AW) saw a small patch at Point Five Gully and a big one at Observatory Gully. Others survived 1951 and 1967 at sites 1, 5–8 and 10–13 in Table 2, and 1951 at site 9. Snow survived 1967 and 1994 near the top of North-East Ridge, and in 2000 below Number Three Buttress and above Tower Scoop (Watson *et al.* 2001). Snow vanished at Zero Gully before that at Point Five Gully or Observatory Gully, as exemplified in Humble's (1946, p. 84) photograph. No snow survived on Ben Nevis in 1945, 1949, 1953 (gone by 13 October), or 1958, 1959, 1969 and 1976 (all so far S. Gordon, as above), 1996–98, 2001, and 2003–06 (Watson *et al.* 1997–2007). Earlier cases of no survival were 1933, 1935, 1938 and 1940 (Manley 1971b).

The year 1908 was unusual, in that Raeburn (1909) saw no snow in Observatory Gully, although patches still occurred below North-East Buttress (i.e. the foot of Zero Gully) and in Coire na Ciste below the Comb. Hence Ben Nevis still held snow, but the patch at the usual most persistent snow-bed had vanished.

East of Ben Nevis stands Carn Mor Dearg, where two patches survived 1967. On Aonach Mor further east, snow survived 1948, 1951, 1956, 1962, 1967–68, 1983, 1988, 1990, 1991 and 1994 (AW), and also in 1999–2000 and 2007–08, but not 1945, 1949, 1953–54, 1957–61, 1969–74, 1976, 1981, 1984, 1987, 1989, 1992, or (Watson *et al.* 1997–2010) 1996–98, 2001–06 and 2009.

At Aonach Mor, big patches survived 1988, 1990 and 1994, and two small ones in 1995. In number and length they exceeded Ben Nevis ones in 1990 and 1994. Each of two surviving patches on Aonach Mor in 1983 was 65 m long, and each of several in 1990 >300 m. On 30 September 1990, several long deep wreaths lay, far bigger than those visible from the road between Inverlochy and Roy Bridge. Patches on Aonach Mor in 1991 totalled 50 m in length. In 1994, three patches totalled 160 m long, including one lying south of Easy Gully at the foot of a climb now named Pirhana. This one, the biggest of the three, was 80 m long.

Aonach Mor's main sites lay in Coire an Lochain, but snow survived also to the south in 1951, 1967, 1990 and 1994, and in 1951, 1967 and 1990 high in the corrie of Allt Dubh to the north (exact sites not noted). After snowstorms with winds from south round by south-west to north-west, cornices form along Coire an Lochain's cliff-top, and in late winter 'can reach monstrous proportions' (Richardson *et al.* 1994). Because all the sites get much sun, however, declines can be rapid. Six patches, each up to 50 m long and totalling 140 m on 5 August 1984, had gone by 30 September, and four up to 50 m on 2 September 1987 vanished by 7 November despite fresh snowfall. Two patches on 17 September 1994 totalled 65 m and a 380-m one carried 10 cm of new snow (D. Scott, photo), but on 29 November the first two had gone, and the big one had broken into three (above). Snow vanished by 7 September 1973 and 1992, and also in 1996–98 by 4, 10 and 1 September. All survival sites in the east corries of Aonach Mor and below the north and east walls of Aonach Beag receive much spindrift and avalanche snow (M. Tighe).

As seen by AW from the Cairngorms, the upper parts of Aonach Beag's north and east faces held surviving snow in the same years as nearby Aonach Mor, except in 1978, 1995 and 2000. In 1984, 20 patches with individual sizes up to 55 m long on Aonach Beag and estimated to attain a total length of 240 m on 5 August had gone by 30 September. Two that I estimated to have a total length of 55 m on 29 November 1994 were still there a few days later, just before lasting snowfall.

Aonach Mor had bigger patches in 16 out of 17 autumns up to 1995 when AW saw both hills clearly on the same autumn day from the Cairngorms, and held surviving snow in three years when he saw none left on Aonach Beag. Only in 1995 did he see it survive on Aonach Beag but not on Aonach Mor. A caution is that most patches at the foot of Aonach Beag's north and east faces are invisible from the Cairngorms. In the experience of resident mountaineers (D.

Scott since 1980, and later I. Sykes, T. Cardwell, S. Fraser and M. Tighe as told to AW), Aonach Mor had more snow than Aonach Beag in most autumns. This is open to some doubt, however, because the most persistent snow-patch at Aonach Beag, described in the next paragraph, is hidden from the main road and most viewpoints.

One of several patches invisible from the Cairngorms is a low-lying hollow at 955 m below Aonach Beag's north wall, remarkable because of its low altitude. Despite being located by Dansey (1919, 1920) and noted (Dansey 1935) in 1933 as lasting longer than the snow at Observatory Gully and possibly longer than the snow on Braeriach, to my knowledge other authors did not mention it until Watson *et al.* (2000). It held a patch on 20 September 1973, shortly before lasting snowfall (J. Duff as told to AW). Aerial photographs (I. Sykes) in autumn 1988 show it as 95 m long (measured on the photograph by AW in comparison with the known height of the cliff) and Aonach Mor's longest one as 49 m (note, however, that photographs were taken well before lasting snowfall, so there would have been further melting).

In 2002, snow survived there but vanished on Aonach Mor. In late October 2002, with lasting winter snow already lying, it was much bigger than the last patch on Ben Nevis at Observatory Gully (Watson *et al.* 2003). It lasted longer in 2004 than on Aonach Mor, though eventually vanishing, and survived in 2007–09 till winter as a larger patch than any on Aonach Mor. By contrast, although it was the biggest patch to survive 2000 on Aonach Beag, a surviving patch on Aonach Mor was twice as long in that year. Annual reports by Watson *et al.* since and including 2000 shows that snow at the 955 m site has been more persistent than at Aonach Mor in 2002, 2003, 2004, and 2007–10, vice versa in 2000 and 2001, and not enough information to make the comparison in 2005 and 2006. In addition, in 1973 and 1988 it was more persistent than on Aonach Mor, so that comes to a total of seven years with surviving snow more than at Aonach Mor, and two less. It seems likely that it has been under-recorded until recent years.

The massif east of Aonach Beag is called The Grey Corries. It includes several high hills, rising to 1177 m at Stob Choire Claurigh, and often carries much snow in early summer, but seldom into autumn. During at least 1951 and 1967, however, snow survived until winter on Stob Coire Easain (AW).

Snow on Beinn a' Chaorainn at Loch Laggan survived in 1951, 1967 and 1994. In August 1994, H. Macinnes used radar to find under the snow the body of a man who had fallen through a cliff-top cornice on 22 January and was not found in prolonged searches by mountain rescue teams (M. Tighe told to AW). The patch, >100 m long in August, became very small in November, but survived.

From the Cairngorms, AW saw snow surviving on Ben Alder near Dalwhinnie in 1947, 1951, 1956, 1967, and 1994, on nearby Geal-charn in these years and also 1974, 1983, 1986, 1990 and 1993, and on the Geal Charn by Ardverikie in 1951 and 1967. He saw none surviving on Ben Alder or the Ardverikie Geal Charn in 1945, 1948–49, 1953–54, 1957–61, 1969–71, 1976, 1980–81, 1985, 1987–89, 1992 and 1995, and none survived 1996–09 (Watson *et al.* 1997–2010). Ben Alder's two most persistent patches are invisible from the Cairngorms, but site visits showed none surviving 1948–49, 1959, 1971 and 1981 (AW), or 1996–2009 (Ian Crichton and Watson *et al.* 1997–2010) or 2010. A patch at Aisre Cham on Geal-charn survived 2000, but not 1996–99 or 2001–09 (Ian Crichton and Watson *et al.* 2001–10) or 2010. G. Oswald, head stalker in 1965–91, told AW in 1997 that snow survived on these hills in eight years out of 10 up to 1980, but then tended to vanish more often.

Snow survived on Sgurr nan Clachan Geala of Glen Cannich, a subsidiary top of Sgurr na Lapaich, in 1951 and 1955 (AW, and sheep-tenant A. Murray as told later to AW), and in 1967 and 1994. Though 25 m long on 20 September 1995, it vanished by 1 November. It survived 1951 also in Coire Beag to the south (A. Murray). On An Riabhachan to the west, snow in Coire Gnada facing Glen Strathfarrar survived 1967 (AW) and lasted till about 1 October 2000 (Watson *et al.* 2001).

In Coire an t-Sneachda on the south side of Toll Creagach at Glen Affric, snow survived 1951 and 1955 (AW, A. Murray), and 1967, 1983 and 1993 (AW, last two years noted also by Pottie 1995). Though only 25 m long on 2 September 1987, it was there on the 27th but had gone by 1 October. Snow survived 1955 and 1967 below the corrie north of the summit of Toll Creagach. On the next hill to the west, Tom a' Choinich, a small patch survived in 1951 and 1955 near the top on its east side (AW, A. Murray), and another in 1967 facing Glen Cannich. A small patch survived 1951 on Carn Eige further to the west, and another on nearby Mam Sodhail.

At Coire nan Con of Ben Wyvis, a patch at the foot of a gully on 4 August 1948 had gone by the 31st. In a hollow above Coire nan Con, snow in 1951 lay into September (AW) and was reported surviving till winter (A. Murray told to AW), while in 1967 it lasted at least into mid September. It vanished there in almost all other years in 1945–69 (AW from the Cairngorms or east Banffshire coast), but the record was incomplete because of poor visibility in a few years. No snow has survived on Ben Wyvis since 1970 (JP, AW, J. Mackenzie told to AW). Snow above Coire nan Con lay till 22 October 1994 (JP), but lasting new snow did not fall till December.

In the warm autumn of 1949, all snow had gone from Ben Wyvis and the hills of Glen Affric, Cannich and Strathfarrar long before 24 September. Stalkers D. Fraser of Inchvuilt and Mr Fleming of Monar told AW in September 1949 that none survived on any of these hills in 1939–49, and D. Maclennan at Affric Lodge (as told to AW in April 1951) confirmed this for 1939–50.

Aspect of survival sites

Out of 96 extra north-east Scottish sites in 1942–70 (Table 1), 50% faced north-east and 39% south-east, but only 3% south-west and 6% north-west. This contrasted with the 69%, 25%, 0%, and 5% of the 55 sites in 1971–2000 (Watson *et al.* 2002). The high proportion facing south-east in the extra 1942–70 sites reflected the mainly north-west winds during the snowfalls of 1951. However, because all of the 1971–2000 sites held surviving snow in 1942–70, the two sets are not independent and statistical analyses comparing them would be invalid.

Of 61 known west-Scottish survival sites, 80% faced north-east, 11% south-east, 3% south-west, and 5% north-west (Table 2), showing a strong north-east predominance. In most years, AW viewed western hills from the Cairngorms to the east, so did not see most slopes facing south-west or north-west. Nevertheless, his viewing from all other directions in 11 years in 1948–67 and all 10 years in the 70s showed no surviving snow that faced south-west, and likewise such viewing annually since 1996 by JP and local observers.

Altitude of survival sites

The 96 extra north-east Scottish sites in 1942–70 (Table 1) spanned a wide altitude (median 1067 m, range 740–1255 m). They tended to be only slightly lower than the 55 sites with surviving snow in 1971–2000 (review by Watson *et al.* 2002). Here I have recalculated the 55 because Iain Cameron in 2009 recorded the altitudes of the most persistent two patches by GPS (Watson *et al.* 2010), each being 5 m higher than given in the review. The amended median is 1100 m. The 61 west-Scottish sites were only slightly lower (median 1005 m, range 725–1300 m). Because of far less effort in the west before 1995, it would be invalid to compare west with north-east sites, except in 1996–2010 when total surveys covered both areas. In these 15 years, the 12 west sites did not differ materially in altitude from the 11 north-east ones (medians 1092 and 1115 m)

Snow that nearly survived

In sites where snow was not seen to survive but lay late, it might survive in a year with early heavy snow. Table 3 lists four such north-east sites and a minimal 10 western ones.

Snow lay in the Corrie of Lochnagar till 27 October 1994, and in 1967 and 1974 till mid October. It did not survive in 1938–2010 there (AW) or on nearby Cairn Bannoch (deerstalkers J. Robertson and J. Robertson snr, told to AW). The Corrie's latest snow lay in the foot of Douglas-Gibson Gully, the next latest below Parallel Gully A.

Snow at Meall a' Bhuiridh near Glencoe did not survive in 1990–2004 (P. Weir and D. Patterson told to AW), or since (I. Cameron).

Acknowledgements

I thank Mark Atkinson, Iain Cameron, Attila Kish, Douglas Scott, Ian Sykes and Mick Tighe for photographs and observations, and for observations Mervyn Browne, Tony Cardwell, Ian Crichton, John Duff, Duncan Fraser, Simon Fraser, Alan Gorham, Peter Hodgkiss, Hamish Macinnes, John Mackenzie, Duncan Maclennan, Alex Murray, George Oswald, Dave Patterson, John Robertson junior and senior, and Peter Weir. Dr Richard Davison and Prof Julian C Mayes read the manuscript and gave valuable advice.

Errata

In Watson, Davison & Pottie (2002), Table 1, location 1 should be at 940981, 2 E of 940981, 4 S of 940981, and snow at location 25 survived 1986 in addition. I estimated altitudes of locations 1 and 2 from map contours, but IC in 2009 measured them by GPS as 1145 and 1140 m (Watson *et al.* 2010), each 5 m higher; results of statistical analyses are unaffected. In the References, for Tabler (1975) the volume number is 43.

References

Cameron-Swan, D. (1900). A snow bridge in Coire ard Dhoire (Corrie Arder). Scottish Mountaineering Club Journal 6, 22–23.

Dansey, R.P. (1919). A tramp between Lochaber and the Cairngorms. Scottish Mountaineering Club Journal 15, 196–206.

Dansey, R.P. (1920). The Aonachs. Scottish Mountaineering Club Journal 16, 192–193.

Dansey, R.P. (1935). Disappearance of the Scottish snow-fields. Scottish Mountaineering Club Journal 20, 366–367.

Humble, B.H. (1946). On Scottish hills. Chapman & Hall: London.

Manley, G. (1952). Climate and the British scene. Collins: London.

Manley, G. (1971a). The mountain snows of Britain. Weather 26, 192–200.

Manley, G. (1971b). Scotland's semi-permanent snows. Weather 26, 458–471.

Pottie, J.M. (1979). Scottish snowbeds in 1976. Weather 34, 81.

Pottie, J.M. (1995). Scottish snowbeds: records from three sites. Weather 50, 124–129.

Raeburn, H. (1909). The North Buttress, Carn Dearg of Nevis. Scottish Mountaineering Club Journal 10, 183–189.

Richardson, S, Walker, A, Clothier, R. (1994). Ben Nevis rock and ice climbs. Climbers' Guide, Scottish Mountaineering Club.

Spink, P.C. (1968). Scottish snowbeds in summer 1967. Weather 23, 209–211.

Spink, P.C. (1980). A summary of summer snow surveys in Scotland: 1965–1978. Journal of Meteorology 5, 105–111.

Watson, A., Cameron, I., Duncan, D. & Pottie, J. (2009). Twelve Scottish snow patches survive until winter 2008/2009. Weather 64, 184–186.

Watson, A., Cameron, I., Duncan, D. & Pottie, J. (2010). Six Scottish snow patches survive until winter 2009/2010. Weather 65, 196–198.

Watson, A., Davison, R.W. & Pottie, J. (2002). Snow patches lasting until winter in north-east Scotland in 1971–2000. Weather 57, 374–385.

Watson, A., Duncan, D., Cameron, I. & Pottie, J. (2008). Nine Scottish snow patches survive until winter 2007/2008. Weather 63, 138–140.

Watson, A., Duncan, D., Cameron, I. & Pottie, J. (2009). Twelve Scottish snow patches survive until winter 2008/2009. Weather 64, 184–186.

Watson, A., Duncan, D. & Pottie, J. (2007). No Scottish snow survives until winter 2006/07. Weather 62, 71–73.

Watson, A., Duncan, D. & Pottie, J. (2006). Two Scottish snow patches survive until winter 2005/06. Weather 61, 132–134.

Watson, A., Pottie, J. & Duncan, D. (1998). Survival of two snow patches in the UK until winter 1997/98. Weather 53, 155–158.

Watson, A., Pottie, J. & Duncan, D. (1999). Only one UK snow patch lasts until winter 1998/99. Weather 54. 369–374.

Watson, A., Pottie, J. & Duncan, D. (2000). Six UK snow patches last until winter 1999/2000. Weather 55, 286-290.

Watson, A., Pottie, J. & Duncan, D. (2001). Forty-one UK snow patches last until winter 2000/01. Weather 56, 404–407.

Watson, A., Pottie, J. & Duncan, D. (2002). Two UK snow patches last until winter 2001/02. Weather 57, 219–222.

Watson, A., Pottie, J. & Duncan, D.(2003). Five UK snow patches last until winter 2002/03. Weather 58, 226–229.

Watson, A., Pottie, J. & Duncan, D. (2004). No Scottish snow patches survive through the summer of 2003. Weather 59, 125–126.

Watson, A., Pottie, J. & Duncan, D. (2005). One Scottish snow patch survives until winter 2004/05. Weather 60, 165–166.

Watson, A,, Pottie, J, Rae, S, Duncan, D. (1997) Melting of all snow patches in the UK by late October 1996. Weather 52, 161.

Wilson, J.W. (1958). Dirt on snow patches. Journal of Ecology 46, 191–198.

Table 1. Location number and grid reference, type (A top of a gully at top of cliff or steep slope, B gully or hollow at foot of cliff or steep slope, C open ground at top of steep slope, D circular hollow on plateau, E long hollow on plateau, F avalanche hollow below cliff or steep slope; G hollow within a cliff or steep slope, H middle of a gully on a cliff or steep slope), year, altitude (m), and aspect where snow survived in north-east Scotland in one or more years in 1942–70, additional to the 55 in 1971–2000 (Watson *et al.* 2002). All were in the Cairngorms massif (grid letters NH, NJ, NN and NO), except 23 on NH, 22, 35 and 73–76 on NN, and 6 and 65 on NO.

Location and grid	Type	Year	Altitude	Aspect	Location and grid	Type	Year	Altitude	Aspect
1, S of 978011	D	47	1080	NW	49, 012042	D	51, 67	1085	SE
2, NW of 983000	D	47	1135	SW	50, W of 990020	E	51, 67	1120	SE
3, N of 985998	E	47	1220	NW	51, S of 945985	B	51, 63, 67	1050	NE
4, S of 000041	C	47	1075	NE	52, 944989	B	51, 67	1050	SE
5, 977019	E	47	1095	NW	53, N of 944989	B	51, 67	1050	SE
6, 205848	D	47	910	NW	54, E of 944988	B	51, 67	1010	SE
7, S of 985027	B	47	1070	NW	55, 905976	B	51, 67	840	NE
8, 987993	E	47, 53	1210	NW	56, E of 009030	E	51, 67	1055	SE
9, NW of 093005	A	51	1170	SE	57, W of 942977	A	51, 67	1160	SE
10, 091000	A	51	1140	NE	58, W of 944976	A	51, 67	1100	NE
11, 094004	B	51	1040	SE	59, E of 945976	A	51, 67	1080	NE
12, E of 091991	B	51	970	SE	60, E of 941978	B	51, 67	1150	NE
13, 104004	E	51	1060	SE	61, E of 940979	A	51, 67	1190	NE
14, N of 086972	C	51	920	SW	62, W of 940979	B	51, 67	1110	SE
15, 057979	C	51	790	NE	63, S of 940979	A	51, 67	1190	NE
16, 024987	B	51	890	SE	64, NE of 160026	B	51, 67	740	NE
17, SW of 025015	D	51	1115	SE	65, 171767	G	51, 67	970	NE

18, NE of 999987	A	51	1100	SE	66, NE of 940942	C	51, 67	1060	NE
19, 001988	B	51	1050	SE	67, E of 941943	G	51, 67	990	NE
20, 002997	E	51	1090	SE	68, NE of 958922	C	51, 67	1075	NE
21, NW of 991986	C	51	1230	SW	69, E of 958921	C	51, 67	1055	NE
22, 934841	C	51	920	NE	70, N of 095009	C	51, 67	1160	NE
23, 633031	E	51	900	SE	71, NW of 113007	A	51, 67	1050	SE
24, S of 157014	C	51	850	SE	72, NW of 113008	A	51, 67	1060	NE
25, E of 954995	A	51	1160	SE	73, NW of 433880	B	51, 67	790	SE
26, 936976	E	51	1110	SE	74, 434875	A	51, 67	1025	NE
27, SE of 968925	E	51	990	NE	75, NE of 422878	A	51, 67	1050	SE
28, S of 898920	C	51	850	SE	76, SE of 430867	A	51, 67	950	NE
29, NW of 900922	C	51	890	SE	77, SW of 938955	C	51, 67	970	NE
30, E of 939980	B	51	1200	NE	78, NW of 940952	C	51, 67	985	NE
31, NE of 940981	A	51	1200	SE	79, 909969	G	51, 67	820	SE
32, S of 964000	A	51	1135	NE	80, W of 990029	B	51, 67	1100	NE
33, E of 952993	A	51	1170	SE	81, 007048	D	51, 67	1090	NE
34, SW of 946992	B	51	1100	SE	82, SE of 014996	A	51, 67	1010	NE
35, SW of 432871	G	51	990	SE	83, 999994	E	51, 67	1100	NE
36, NW of 887925	C	51	930	SE	84, W of 004997	E	51, 67	1050	NE
37, SE of 998983	G	51	1115	SE	85, 002997	E	51, 67	1120	SE
38, 084987	E	51, 67	1050	SE	86, SE of 999991	E	51, 67	1150	NE
39, 110012	A	51, 67	1065	NE	87, S of 002000	E	51, 67	1100	NE
40, 999989	E	51, 67	1185	NE	88, NW of 003005	E	51, 67	1070	NE
41, SE of 999991	D	51, 67	1180	NE	89, 944935	E	51, 67	1050	NE

42, S of 002000	B	51, 67	1080	SE	90, 944937	E	51, 67	1040	NE
43, NW of 003005	A	51, 67	1060	NE	91, 993942	E	51, 67	980	NE
44, SE of 002013	B	51, 67	945	NE	92, 944973	E	51, 67	1050	SE
45, N of 996000	E	51, 67	1125	NE	93, 950970	E	51, 67	1050	SE
46, SW of 996992	E	51, 67	1090	NE	94, 958973	A	51, 67	1100	NE
47, S of 994996	E	51, 67	1200	NE	95, 136993	A	51, 67	930	SE
48, 993996	E	51, 67	1255	NE	96, 105015	A	67	1110	NE

1–2 March Burn, 3 & 8 Coire Mor, 4 Coire Cas, 5 Miadan Creag an Leth-choin, 6 Carn an t-Sagairt, 7 Coire an Lochain, 9–11 Coire nan Clach, 12 Coire an Dubh-lochain, 13 Allt Dearg, 14 Coire Gorm, 15 Allt Clais nam Balgair, 16 Coire an Lochain Uaine of Derry Cairngorm, 17 Beinn Mheadhoin, 18–19 Coire Sputan Dearg, 20 & 41–43 & 83–88 Loch Etchachan, 21 Coire Clach nan Taillear, 22 An Sgarsoch, 23 Carn Ban, 24 Allt Phouple, 25 Coire Bhrochain, 26 SW of Garbh Choire Mor, 27 Coire Caochan Roibidh, 28–29 Coire Eindart, 30–31 and 57–63 Garbh Choire Mor, 32 Coire na Lairige, 33 W of Coire Bhrochain, 34 & 51–54 Garbh Choire Dhaidh, 35 & 73–76 Creag Meagaidh, 36 Coire Bhlair, 37 Coire an Lochain Uaine of Ben Macdui, 38 Alltan na Beinne, 39 & 96 Garbh Choire of Beinn a' Bhuird, 40 N of Coire Sputan Dearg, 44 Castlegates Gully, 45–48 upper Garbh Uisge Mor, 49 S of Ciste Mhearad, 50 N of Feith Buidhe, 55 Coire Odhar nan Each, 56 Stac an Fharaidh, 64 Feith Laoigh, 65 Brudhach Mor, 66–67 Coire Creagach, 68–69 Coire an t-Sneachda of Beinn Bhrotain, 70 Feith Ghiubhasachain, 71–72 Cnap a' Chleirich, 77–78 Monadh Mor N, 79 Coire Odhar, 80 Coire an t-Sneachda of Cairn Gorm, 81 Coire na Ciste. 82 Creagan a' Choire Etchachan, 89–91 Leac Ghorm, 92 Clais Luineag, 93 Allt Clais an t-Sabhail, 94 Coire an Lochain Uaine of Cairn Toul, 95 Allt an Aitinn.

On Creag Meagaidh, locations 35 & 76 were in Coire Choille-rais, and 73 was at the foot of Easy Gully in Coire Ardair, where a September snow-tunnel cut by a stream is in a photograph by Cameron-Swan (1900), 74 was at the top of Easy Gully, and 75 in the corrie SW of Uinneag Coire Ardair. Other snow patches survived 1951 and 1967 on Creag Meagaidh in Coire Choille-rais and the top of Raeburn's Gully in Coire Ardair, and in 1967 at the corrie N of the summit, but exact locations not noted; likely places, to judge from snow lying long in recent years, are SE of 434875 and 1030 m for Raeburn's, 416876 at 1070 m and 407872 at 980 m for northern ones, and all three facing NE.

Extra unlisted patches survived 1951 in Coire an Dubh-lochain, Coire nan Clach and Garbh Choire of Beinn a' Bhuird, on Ben Macdui (especially at upper Garbh Uisge Mor and above Loch Etchachan), on Braeriach (especially Coire Bhrochain and Garbh Choire Dhaidh), Monadh Mor's E corries, and Beinn Bhrotain's SE corries, but exact sites were not noted. Patches at the end of September lay below Hell's Lum Crag at Loch Avon and below Crimson Slabs in Coire Etchachan, so snow may have survived there. It lay till late September 1951 at 800 m in Coire Gorm of Beinn Bhreac, and at 780 m in Coire Gorm of Beinn a' Bhuird, but not till winter. Snow at unlisted locations also survived 1947, 1963 and 1967, but exact sites were not recorded.

Table 2. Location number and grid reference (grid letters NN at locations 1–51, NH at 52–60), type (A etc as in Table 1), altitude (m) and aspect of west-Scottish patches known to survive in one or more years in 1945–2010.

Location	Type	Altitude	Aspect	Site	Type	Altitude	Aspect
1, SW of 169716	B	955	NE	31, NE of 196731	F	1045	NE
2, 167715	B	1095	NE	32, NE of 195723	B	995	SE
3, 166714	B	1130	NE	33, SE of 192739	B	1130	NE
4, S of 166713	H	1205	NE	34, 236729	E	980	SE
5, 165713	B	1250	NE	35, 201653	B	940	SW
6, W of 165713	A	1300	NE	36, S of 202653	B	965	SW
7, E of 162714	B	1160	NE	37, NE of 156535	B	940	NE
8, NW of 161715	B	1095	NE	38, NW of 496714	G	1050	SE
9, E of 162718	F	910	NE	39, NW of 496729	D	945	NE
10, E of 169714	C	1300	NE	40, SE of 485722	D	1045	NE
11, NW of 159717	B	1115	SE	41, NE of 491728	A	995	NE
12, SE of 164715	B	1190	NW	42, NW of 483745	G	945	NE
13, E of 170715	G	1230	NE	43, SW of 479745	C	1070	SE
14, NW of 182725	F	860	NE	44, NE of 481746	B	1005	NE
15, E of 180723	F	1000	NE	45, NE of 480748	G	1030	NE
16, 203716	F	725	NE	46, SE of 483745	E	930	NE
17, 197718	B	955	NW	47, N of 486745	B	900	NE
18, NE of 199714	G	1000	NE	48, 507812	A	980	NE
19, 198715	G	1155	NE	49, SE of 387852	F	950	NE
20, NE of 197718	B	1005	NW	50, 388849	F	900	NE
21, E of 200715	G	940	NE	51, 387848	F	920	NE
22, NW of 200710	B	1080	NE	52, 162344	G	1015	NE
23, NE of 200712	G	985	NE	53, NE of 142351	G	870	NE
24, N of 200710	B	1050	NW	54, NE of 193284	B	940	NE
25, 194737	F	1090	NE	55, E of 197278	B	890	NE
26, SE of 194736	F	1130	NE	56, NE of 197279	B	900	SE
27, 193737	B	1120	NE	57, E of 164273	B	1090	NE
28, N of 194738	F	1015	NE	58, 165274	C	1030	NE
29, SW of 193737	F	1140	NE	59, 131263	B	1050	SE
30, NW of 196730	B	1125	NE	60, 121255	G	1050	NE
				61, S of 466687	E	965	NE

1–13 Ben Nevis (1 foot of Zero Gully, 2 foot of Point Five Gully, 3 top of Observatory Gully, 4 Gardyloo Gully, 5 shelf above Tower Scoop, 6 Tower Gully, 7 foot of Number Two Gully and foot of Comb Gully, 8 under Number Three Gully Buttress, 9 Lochan Coire na Ciste, 10 top of North-East Ridge, 11 Number 4 Gully, 12 Garadh na Ciste and Glover's Chimney, 13 shelf on North-East Buttress), 14–15 Carn Mor Dearg, 16–24 Aonach Beag (17 main site), 25–33 Aonach Mor (25 protalus rampart, 27 main cliff-foot site), 34 Stob Coire Easain, 35–36 Na Gruagaichean, 37 Stob Coire Sgreamhach, 38–41 Ben Alder, 42–47 Geal-charn, 48 Geal Charn, 49–51 Beinn a' Chaorainn at Moy, 52 Sgurr nan Clachan Geala, 53 An Riabhachan, 54–56 Toll Creagach, 57–58 Tom a' Choinich, 59 Carn Eige, 60 Mam Sodhail, 61 Ben Wyvis. Cameron of Locheil called 3 The Catskin from its shape (to David Laird about 1963, told by DL to AW in 2003), well shown in Fig. 2 of Watson *et al*. (1999).

Table 3. Location number and grid reference (grid letters NN at 7, 10–11 and 14, NO at 6 and 8-9, others NH), year, time of vanishing, type (A etc as in Table 1 title), altitude (m) and aspect of late patches that did not survive but might have survived if heavy drifting snow had come early, as in September 1976 in the Cairngorms.

Location and grid	Year	Time of vanishing	Type	Altitude	Aspect
1, NE of 130261	51	early October	B	1040	SE
2, SW of 121258	51	early October	B	990	SE
3, 983000	51	early October	E	1130	SW
4, 291424	51	start of September	E	900	SE
5, NE of 294424	51	start of September	B	850	SE
6, SW of 128455	67	late September	G	910	NE
7, W of 906995	67	end September	A	1140	SE
8, 004060	72	end October	H	820	NW
9, 143544	82	end October–start November	B	980	NE
10, 223828	94	25 September	E	920	NE
11, 248856	94	27 October	B	950	NE
12, 887970	94	6 October	E	930	NW
13, E of 257507	94	early September	B	880	NE
14, 503712	94	12 October	A	770	NE
15, 462679	94	24 October	B	690	SE
16, 415383	notes	September	D	1030	SE

1 Carn Eige, 2 Mam Sodhail, 3, 5 Sgurr na Ruaidhe, 4 Ben Macdui, 6 Maoile Lunndaidh, 7 Sgor Gaoith, 8 Coire na Ciste of Cairn Gorm, 9 Bidean nam Bian, 10 Cairn Bannoch, 11 Corrie of Lochnagar, 12 Ciste Mhearad of Feshie, 13 Meall a' Bhuiridh, 14, 15 Ben Wyvis, 16 Beinn Heasgarnich. Snow vanished at site 9 in Central Gully of Bidean nam Bian before lasting snowfall at the end of November (H. Macinnes letter to AW). Location 13 was at the foot of Flypaper ski-run (P. Weir shown to AW). M. Browne (letter to AW) saw snow at location 16 in September of a few years but not surviving except maybe in 1951.

Chapter 6. Cairngorms glaciers in the 18th and 19th centuries highly unlikely

Historical evidence on snowier conditions in the 1700s and 1800s during the Little Ice Age can be compared with recent research on the time that has elapsed since the ground became free of ice or perennial snow. The lichen *Rhizocarpon geographicum* grows on cliffs and boulders, in the form of roughly circular patches that increase in diameter year by year as the lichen grows, and the growth rate has been measured in several countries. Sugden (1974, 1977) measured the diameter of the biggest patches of this lichen in seven corries in the Cairngorms, and compared this with other workers' measurements of the growth rate. He and his student Sheila Rapson found that the diameter on the downhill side of the outermost moraine exceeded that on the uphill side of the innermost moraine. He inferred that the biggest (i.e. oldest) of the lichen patches on the uphill side of the innermost moraine had started to grow between 1644 and 1725, and much more recently about 1800 on an innermost ridge much nearer the present Garbh Choire Mor snow-bed. Though postulating (1977, Conclusion) that glaciers occurred there prior to these years, he referred several times in his main text to 'snow or ice'. Lamb (1977) recounts Manley's suggestion to him 'that it may have been firn rather than having attained the density of ice'. This fits observations of no lichen on bedrock and boulders beside firn, as well as beside glaciers.

Because no lichens grow on boulders, bedrock or subsoil at today's two most persistent snow patches at Garbh Choire Mor (above), one need not even postulate firn. Long-lying snow would suffice. No rock lichens grow today on boulders at snow patches of far less persistence than the two at Garbh Choire Mor, e.g. at Ciste Mhearad and Garbh Uisge Beag. However, it follows from Sugden's study that long-lying snow must have covered a far bigger area than in recent decades. In 1943–2010, no snow survived immediately behind the inner moraines in any of his seven corries apart from Garbh Choire Mor, and none immediately behind inner moraines in any other corrie in the Cairngorms.

Later research by radiocarbon dating and pollen analysis shows that the moraines formed more than 6000 years ago, not in recent centuries (Rapson 1990). Noting that 'A quick glance at the area immediately surrounding the semipermanent snowpatch in Garbh Choire Mor today clearly shows the ability of snow to kill lichens', Rapson suggested that extra snow inside the moraines during the Little Ice Age would have killed rock lichens, thus explaining their current smaller diameter inside the moraines than outside them. However, she wrote that the inner moraine at Garbh Choire Mor may be an exception. Although it lacked enough organic material for tests by radiocarbon dating and pollen analysis, she commented on the moraine's current fresh appearance, with hardly any vegetation and with boulders angular and unweathered. She suggested that a glacier in the Little Ice Age may have formed the inner moraine.

An alternative and far more likely explanation is that the ridge is not a moraine but a protalus rampart. Each year, boulders and soil becoming loose and falling from the cliffs, or being torn out by snow avalanches, slide down the steep surface of the snowfield immediately below the crags. They then pile up far below, adding to the ridge. On several days from October through to early summer of different years I have watched this happening, and the signs of it are clearly evident on many days in every summer. Although Berry (1967) did refer to the moraine lip of the upper corrie, he then added, 'This is probably a type of avalanche boulder tongue, built up by the accumulation of avalanche debris and by the rocks released from the walls of the corrie by frost shatter'. I agree with his suggestion in the last quoted sentence.

The angular boulders mentioned by Rapson would be unlikely to have resulted from deposition in the form of a moraine, because glacial action would have rounded them. Hence I suggest the hypothesis that the ridge is a protalus rampart. If boulders below the surface are angular and the soil contains little clay or silt, this would tend towards refuting the hypothesis of the ridge being a moraine, and support its being a protalus rampart. This can be put to the test by detailed inspection of boulders, soils and vegetation, using the well-tested methods of soil survey, It seems suprising that this was omitted by geomorphologists..

The mound lies in the corrie's centre below the long-lying snow, south of 943 980, with a slight hollow on its west side. Sugden did not inspect the type of rocks and soil on the mound. If it is a moraine, rocks would be rounded due to grinding by ice, and all exposed faces of surface rocks would carry some lichen unless there was a long-lying snow patch nearby. By contrast, a protalus rampart consists largely of angular or sharp-edged rocks that have fallen due to

Garbh Choire Mor avalanche tracks funnel towards boulder ridge on left, 12 June 1948

Garbh Choire Mor avalanche debris funnels to boulder ridge, start of June 1955 (Tom Weir)

Garbh Choire Mor, lower big snow patch from avalanches is held by boulder ridge in centre of corrie, 12 August 1975

frost-shattering of cliffs. Only the outer surface of a cliff has rock-lichen. If a piece breaks off, only its outer face has rock-lichen. If a piece has come from inside the cliff after an outer surface comes off, it will have no lichen on any face. So, if any freshly-fallen rock on the mound has a face with no lichen, this indicates a protalus rampart.

From a distance the mound seems to consist of nothing but boulders, but if no soil is visible on the surface it would be worth looking under a few of the topmost boulders to see if they are lying on soil. If there is any, it is likely to be dark topsoil that has been torn from the cliffs in avalanches.

The almost fluorescent yellow-green of *Rhizocarpon geographicum* imparts a unique hue to the cliffs of Garbh Choire Mor, greater than any other Scottish cliff. It likes snow, but too much snow deters it and other rock-lichen, so the cliff immediately above the longest-lying snow patches shows the pink colour of the local granite.

When the snow patches are tiny or have vanished, it would be useful to record by photos how far the vegetation-free zone extends on boulders and soil around the Sphinx and Pinnacles patches. Lichens, moss and other plants would begin to extend on to this zone if the average duration of snow declines. Photos showing the edge of the zone would be valuable, along with a note showing where the photograph was taken so that the location can be revisited. GPS would be a useful additional tool for this.

On 23 September 2010, Attila Kish visited the snow patches further up the corrie 'in truly awful weather', and on the way to them he inspected the ridge or mound after I had sent him the background information and suggestions about the best aims on a brief visit. He noted that the western part of the mound, facing the upper corrie, consisted of boulders, but the eastern section had a surface with some soil mixed with boulders. On that eastern section he used a trowel to dig a soil pit about 15–20 m downhill from the crest, and took photographs of the pit, using the trowel to give scale. The weather was bad and the light poor, but his preliminary reconnaissance proved useful.

The upper soil horizons comprised '12–15 inches of organic soil before getting to a lighter, what looked like mixed organic/mineral layer. The overlying vegetation was crowberry/Rhacomitrium with alpine club moss and stiff sedge. My memory of soil types is vague, but it certainly wasn't what I think of as classic moraine type soil i.e. a thin layer of organic before getting into the mineral clay/gravel mix. I'm just thinking of 'borrow pits' dug by estates or path contractors in moraine country, where they're getting very quickly through to the material they want for surfacing'.

His photographs confirm his general description above. The thickness of the upper soil horizons above the mineral horizons goes against the hypothesis of a moraine, especially given the exposure and short plant growth at that altitude and location. During avalanches, the larger boulders tend to pile up on the western side of the mound and the finer soil particles and bits of vegetation move further downhill, tending to concentrate on the more sheltered eastern side of the mound. This helps explain the remarkable thickness of the upper horizons. At such an exposed site, an upper horizon of less than an inch would be expected, far less 12–15 inches. Attila Kish intends to revisit the mound in better weather and dig another pit, as well as make observations on the angularity and weathering of the boulders and on whether recently deposited boulders and pieces of rock from the cliff as a result of the previous winter's avalanches are lying on top of boulders with rock lichen or on top of vegetation.

References

Rapson, S.C. (1990). The age of the Cairngorm coire moraines. Scottish Mountaineering Club Journal 34, 457–463.

Sugden, D. E. (1974). Deglaciation of the Cairngorms and its wider implications. Problems of the deglaciation of Scotland (ed. by C.J. Casseldine & W.A. Mirtchell), 17–28. STAG Special Publication 1. Geography Department, University of St Andrews.

Sugden, D. (1977). Did glaciers form in the Cairngorms in the 17th–19th centuries? Cairngorm Club Journal 97, 189–201.

Chapter 7. Vantage points for snow patches at roads in north-east Scotland

Grantown-Tomintoul A939 road, above Corriechullie, layby at 021467. Good view of Coire an Lochain of Braeriach, Coire an Lochain of Cairn Gorm, W part of Coire an t-Sneachda, and shallow corrie N of Cnap Coire na Spreidhe.

Same road, about half way between above layby and Dirdhu, possible to get car just off the road at one spot on S side of road. It is a long straight bit of road, so you can easily see if vehicles are coming. Can see rocky slopes above Loch Etchachan and shallow corrie at top of burn running to Loch Etchachan from N of the Coire Sputan Dearg cliffs.

Same road, further E, layby on S side of road at 104201, can see snow on Sgoran Dubh and Sgor Gaoith.

Same road, car park on left just before sewage works at NW end of Tomintoul. Walk short distance up steep brae on the dirt road in a pinewood above the public road. Good for seeing snow in Lochan nan Gabhar N corrie of Ben Avon and slopes E of it.

Go out of Tomintoul towards the Lecht, but continue straight on past the road to the Lecht and along the B9008 road to Dufftown, over the bridge and through the plantation, to come out beside a peat moss on the left. Continue to the end of the farm road to Inchnacape (signposted), and the entrance is wide enough for you to draw the car in safely off the public road. An excellent place for viewing the Slochd Mor, the slabby corrie on the E side of Cnap a' Chleirich, and the northern corries of Beinn a' Bhuird.

A939 Grantown-Forres road, near Cottartown N of Grantown, can draw in safely at road ends and at a phone box. Good view of NW corries of Beinn a' Bhuird and an unusual view of Sgor an Lochain Uaine through the V of the Lairig Ghru.

A95 Aviemore-Grantown road, near Balnacruie W of the Heather Centre and S of Croftjames, easy to get car safely off the main road. Good place for seeing snow on Bynack More, and can see into Coire Dearg of A' Choinneach, a steep rocky corrie that often has snow at the start of July.

As above, near Deishar NW of Boat of Garten, fast traffic, so best to take car into the end of the road to Deishar. Can see into Coire Odhar at top of Loch Einich and also corries on E side of loch. Can also see these in nearer view from road from Coylumbridge to Glenmore. E of the long layby of former road on the S side of the road at 932100, continue along the road a short distance till you come to a part with some moorland on the right, offering the same view as above. Can get car safely off the road at one point.

Further up the road to Glenmore, the road to Badaguish heads off to the left. You can get a car off the road if you pass the entrance and immediately afterwards stop, on the left side of the road. A better bet is to turn left up the Badaguish road and immediately go left into a small bay off the road. From the side of the main road here you get a good view into the corries S of Sgor Gaoith above Loch Einich.

N end of Loch Pityoulish, can get car off road on N side. Good view of Coire Dhondail and Coire nan Clach to E of Loch Einich.

S of Loch Pityoulish towards Coylumbridge, good view of northern corries of Braeriach, especially of Coire an Lochain. Can get car off road on E side at entrance road to Guislich and also on W side near there.

Mick Tighe found a good vantage point for the longest-lying snow patch at 196720 on Aonach Beag. This site is hidden from most places on low ground, but he discovered a place at an accessible footpath near Roy Bridge at grid reference 294818.

AW with pointer Solitaire skis on Foveran sands by the North Sea, January 1963 (Adam Watson senior)

Chapter 8. Skiing in and near Aberdeen city in the early 1950s

While a student at Aberdeen University in 1949–52 I never went on ski from lodgings at Broomhill Road and Clifton Road to Marischal College, because the Council kept Union Street and Great Northern Road and their pavements snow-free, but I often skied in evenings on streets and pavements. For two winters I had digs at Clifton Road, a street which afforded a good fast descent on hard snow or ice along pavements and occasionally the road. It was hazardous, because even steel edges did not bite well on hard rough ice. My Welsh digs mate Bill Jenkins, learning to ski, suffered a few black and blue bruises from falling on icy pavements. The skis made a loud rough noise, unlike the smooth swish on hard-packed powder. To control the skis was a strain on muscles, but maybe good practice for skiing on very hard icy snow.

He and I went to hospital grounds for skiing in evenings on powder snow. On one evening after a thaw the surface had frozen into a thick hard sheet, and each step made a loud bang like a gun as the ski broke the crust when we climbed uphill. A caretaker heard the bangs and confronted us, somewhat charily. Because the diffuse light from nearby streetlamps was good enough for us to see without torches, and he heard the noises but saw no torch-lights, he'd thought we were burglars! He ordered us to leave, on pain of informing the police. So we left, but could not restrain ourselves from bursting out laughing as soon as we reached the street.

Aberdeen Council did not keep the less busy pavements and roads free of ice and snow as in recent decades. Vero C. Wynne-Edwards, Regius Professor of Natural History who for had been at McGill University in Montreal, lived at 70 High Street, Old Aberdeen. I recall him skiing to Marischal on a few mornings of deep snow. He often skied miles across fields near Aberdeen, especially past Cairncry Road towards Kingswells on that high land. He and I skied occasionally along beaches at Aberdeen, Blackdog and Balmedie, and I often on beaches at Newburgh and Foveran to Menie.

I once joined him on a tour from Mastrick to Brimmond Hill and return by a different route. We crossed scores of fences, often with barbed wire, and even more drystane dykes, because these predominated in that area of very stony soils of the Countesswells Association. Fewer fences and dykes occur now, because farmers have demolished many to enlarge their fields.

Since then I have often skied on farmland and moorland with dykes or fences. At a dyke or fence you cross it or head for a gate if one is nearby. It's easy to open some gates. Others are tied with so much wire or twine that you can cross the dyke or fence sooner. At a dyke, stand with skis parallel to the dyke, then lean on the upper part of the dyke and swing both skis round to face the opposite direction while swivelling your body on the dyke top, using sticks for balance.

A stock fence is easy to cross without removing skis. At a barbed wire fence, there's a risk of tearing your breeks or worse, but a rucksack laid on the top wire does the trick. You stand at a level spot with both skis parallel to the fence. Using one stick for balance, you kick the ski nearer to the fence up in the air and over the fence, and down to face the opposite direction, i.e. an extended kick turn. Then bring the other ski round, and off you go! You use sticks in unison, to keep balance. After a few fences it's easy!

Old stock fences and deer fences have plain wires below the top strand, so you can crawl through while keeping your skis on. You remove your rucksack and then crouch, turning your body to face at right angles to the fence. Next you slide one ski and leg between the wires, hanging on to higher wires for balance, move your body through, and then the other leg and ski. Modern deer fences of wire netting are far worse, reducing access to walkers and skiers, and killing and maiming far more birds than old-style fencing.

Chapter 9. The remarkable snowstorm of early September 1976

Snow patches melted rapidly in a heat wave in August 1976, and Hudson (19977a) reported 19 days with maximum temperatures of 20C or more at Inverdruie in Strathspey. On 31 August, another hot day when I walked to Ben Macdui, only two snow patches remained in north-east Scotland, both in Garbh Choire Mor, and both very small. I thought that if the hot weather continued the larger but shallower one (below Pinnacles Buttress) would not last beyond a week.

However, fresh snow fell on the high Cairngorms by 22.00 on 1 September and 'there was a fair covering the next morning at altitudes above 1080 m' (Hudson 1977b). The next day was cold also. Despite milder weather on 3 and 4 September, the early snowfalls left many small patches on the high Cairngorms and a cornice rim along Garbh Choire Mor. These melted considerably during the day on the 5th. Late on the 6th there was more fresh snow, and heavy snow fell all day on 7 September accompanied by a north gale, with 4 inches of wet snow lying at the Lecht and the road summit of the Cairnwell pass. On Wednesday 8 September, snow fell in a few light showers, and the air was milder, with some thawing, blinks of sunshine and little wind. On that evening the weather turned cold and fresh snow fell on the high hills. Next day, 9 September, I noted a north gale and two inches of rain at Crathes in lower Deeside, 'all snow high up', so this would have been equivalent to about two feet of snowfall. Hudson (1977b) reported that fresh snow on 9 September was deposited down to 490 m on lee slopes due to drifting, and to 580 m on windward slopes.

On 10 September, deep snow covered the high Cairngorms but warmer air had arrived on a north-east wind and the snow had started to thaw on the highest tops. My father went to Cairn Gorm that day and reported snow showers and a bitter wind. He walked from the top chairlift round to the hollow of Ciste Mhearad to the east, and wrote in

New snow, corries of Cairn Gorm from Glen More, 10 September 1976 (Adam Watson senior)

his diary that it was well filled with new drifted snow. His diary also holds his comment that his Norwegian wooden cross-country skis would have been useful. He took some remarkable photographs showing exceptional snow cover. My father told me that he found fresh drifts up to 10 feet deep in hollows on the north and east and south sides of Cairn Gorm, and reported deep snow lying all the way down to Loch Avon. A party of Belgian tourists, equipped for summer conditions, narrowly escaped with their lives on 9 September after a two-day visit to Loch Avon, and 'had to be rescued. They were suffering from exposure. If the party had not chanced to meet two experienced mountaineers, at least three would have died' (Hudson 1977b).

My father and I climbed Morrone on 11 September, and I noted heavy snow on the Cairngorms, especially on Cnap a' Chleirich and the south side of Braeriach. Heavy snow showers continued on the next day. Hudson (1977b) estimated that the total snowfall from the 9th to the 13th was 0.84 m above 1000 m, with drifts exceeding 3 m deep. 'Conditions were such that skiing and ski touring were enjoyed, and skiable snow lay until the 18th', and 'temperatures remained below average until 17 September' (Hudson 1977b).

On 17 September, there were still deep extensive drifts in many places despite warm weather with sunshine. The whole of the southern slope of Cairn Gorm lay under a deep blanket of consolidated packed drifted snow, which had blown in the gale off the north side. I was on Cairn Gorm plateau on Sunday 19th with my father, a day with bright sunshine and strong glare off the fresh snow. Deep snow drifts lay on the plateau and on the corries of Braeriach, completely covering the old snow in Garbh Choire Mor. I wrote that the heavy melting and hot sun made it feel like early June. The south side of Cairn Gorm still had a complete cover of deep drifted snow up to 3 m in depth, and likewise some hollows on Cairn Gorm such as Ciste Mhearad. I wished that we had brought our touring skis, for the snow had a firm uniform surface and skking would have been much easier than walking.

After much thawing in mild weather, by early October most of the fresh snow had gone. A fresh snowfall occurred on 13 October. On 14 October, John Duff measured the snow patches at Garbh Choire Mor. Four patches with a total

New snow, Cnap Coire na Spreidhe of Cairn Gorm, 10 September 1976 (Adam Watson senior)

Loch Avon from Cairn Gorm, 10 September 1976, Lochnagar distant (Adam Watson senior)

length of almost 200 feet comprised new snow, completely burying any old snow. On 17 October from near Braemar I saw only three tiny patches of new snow on Beinn a' Bhuird and three likewise on Coire Clach nan Taillear of Ben Macdui. I saw no others from Derry Cairngorm, apart from the fresh snow wreaths in Garbh Choire Mor.

References

Hudson, I.C. (1977a). Cairngorm snowfield report 1976. Journal of Meteorology 2, 163–166.

Hudson, I.C. (1977b). Record early cold spell in the Scottish Highlands, 9–16 September 1976. Journal of Meteorology 2, 37–39.

Looking west past Cairn Gorm chairlift, 10 September 1976 (Adam Watson senior)

Ciste Mhearad on 10 September 1976 (Adam Watson senior)

Remains of early September snowfall, north-east from Ptarmigan Restaurant, 18 September 1976 (Adam Watson senior)

AW south of Cairn Gorm, 19 September 1976 (Adam Watson senior)

Dusting of new snow persists at Coire Domhain on cold ground where old snow had been, 9 November 1975

Snow dusts Garbh Choire Mor below a fresh cornice, 13 October 1971

New snowfall, Beinn Bhreac and lower Beinn a' Bhuird from west of Braemar, 11 September 1976

Chapter 10. Polygonal hollows and dirt on snow surfaces

The polygonal hollows that develop on melting snow patches have long been of scientific interest, as for example in the References below. The hollows are typically about 1 to 1.5 inches deep (Richardson 1954), this being the maximum depth in the centre of the hollow. Dirt deposits become concentrated along the edges of these polygons, and especially at the points where two or more polygons meet, and photographs in the papers noted below show the hollows and the dirt deposits well. Warren Wilson (1958) noticed on Jan Mayen that a thin deposit of dirt increases the melting rate of snow underneath, causing a depression in the snow surface just over a centimetre deep within six hours of sunny weather. By contrast, a thick deposit reduces melting, so that the snow under it becomes raised above the level of nearby dirt-free snow, sometimes even for more than a metre above. He suggested that the critical thickness for sandy dirt at Jan Mayen may be roughly 3 mm. Most of the text in the papers named in the References is concerned with possible mechanism by which the polygons are formed and by which the dirt gathers and is concentrated. Richardson (1954) observed that the polygonal hollows and dirt lines occurred on the under surfaces of snow patches where large stone blocks had increased melting and led to a gap between the under side of the snow patch and the ground. He also wrote:- 'Although familiar with most British mountain summits the writer cannot recall having seen dirt polygons before finding them on Crossfell in 1951. Mr S.E. Ashmore, who frequently visits Yr Ffos Ddyfan on Carned Llewellyn assures me that he has never seen them. However, Mr Adam Watson recently sent a photograph (see Plate 15 (ii)) of a late-lying snowbed in Garbh Coire, on Braeriach in the Cairngorms. Even from a distance the dirt cappings can be clearly seen. The snowbed was about 28 ft by 10 ft and judging from the photograph its maximum depth might be 2 to 3 ft. Mr Watson claims the feature to be of common occurrence on the Scottish snowbeds.'

Polygonal hollows on underside of snow patch, March Burn, 29 September 1972 (Adam Watson senior)

Dave Sergeant and AW at March Burn, polygonal hollows on underside of snow patch, 30 June 1947 (George Edwards)

Polygonal hollows on vertical side of snow patch, March Burn, 29 September 1972 (Adam Watson senior)

Hollows and dirt lines on steep snow, Coire nan Clach, Beinn a' Bhuird, 10 August 1975

Dirt lines at shallow gradient, Miadan Creag an Leth-choin, 30 May 1982

AW senior on broad snow-ridge caused by wind-blown dead grass reducing the melt rate, Alltan na Beinne, Beinn a' Bhuird, 10 August 1975

AW at apex of narrow snow-ridge holds a tuft of the thick thatch of dead grass (Adam Watson senior)

Polygonal hollows at Ciste Mhearad of Cairn Gorm, 12 September 1975 (Adam Watson senior)

AW senior stands on icy snow on Ben Macdui, looking west past Sgor an Lochain Uaine, small hollows all over a big snowfield, 30 May 1955

Dirt on Cuidhe Crom (curved wreath) of Coire Cas on Cairn Gorm, after winds blew dust from heavily trampled bare ground nearby, tracks on right show where climbers glissaded, 3 July 1983

Polygonal hollows and sorted dirt on snow patch below Pinnacles Buttress of Garbh Choire Mor, 17 October 1953 (Adam Watson senior)

Dirt almost covers snow at trampled ground above Coire an t-Sneachda, 15 May 1982

In my experience, strong winds increase the amounts of dirt on snow patches. The amounts are unusually large immediately after strong winds where snow patches are lying in sites with increased bared ground and loosened soils due to trampling by many people. A good example is at the Cuidhe Crom or curved wreath at the top of Coire Cas on Cairn Gorm.

Richardson (1954) stated:- 'The polygons are small hollows....on the surface of small snowbeds, occurring at times when the snow is 5 to 7 months old during the last few days before final ablation'. In fact the polygons occur on snow surfaces of variable extent, from small snow patches less than 1 m across up to continuous deep snow stretching for miles. They are universal on the upper surfaces of melting snow in mild weather, and likewise on the under surfaces where wreaths have been undercut by streams or avalanches or differential melting beside rocks, or where people make snow holes for shelter. This includes vertical faces, as noted also by Richardson & Harper(1957), who pointed out another universal feature, that they occur on flat snow as well as on upper and under surfaces of varying steepness up to vertical. I have noticed that on continuous deep consolidated snowfields the hollows tend to be smaller in diameter, also less polygonal and more circular. This may be a result of lower air temperatures owing to the extensive snow and also the large amount of reflection of solar radiation back from the white surface.

In mild weather the hollows develop overnight or within a daytime, and certainly in as little as six hours on small snow patches. If the ground and the snow are frozen and the air is calm or with very light wind, typically in a temperature inversion, the hollows do not develop even at air temperatures up to 6 or 7C, especially in sunny weather, but can form rapidly within a few hours in mild winds of 7C or above. I have not seen them develop in Föhn conditions with air temperature below or near freezing and with very dry wind and extremely low humidity. John Pottie (by telephone and

Metal-framed rucksack for scale on snow thatched with wind-blown dead grass at Alltan na Beinne

email) has confirmed to me that he has seen them develop on the upper and under surfaces of a snow patch overnight when he had made a snow hole along with colleagues for spending a night in the Cairngorms.

Warren Wilson (1958) made an important contribution about the organic and chemical contents of the dirt, and found that much of it comprises plant remains. In the Cairngorms, I have noticed a massive increase of dirt on snow patches, and especially plant debris on snow patches, beginning in late May and June. This coincides with the initiation of new bright green plant growth. As the plants grow, the new growth breaks free from seeds and dead leaves, which become separated from the plant and fall to the ground. There they are readily moved by wind.

In this chapter I present several photographs of the sometimes intricate patterns of polygonal hollows and dirt lines. At times the thatch of dead plant litter is so thick that the snow underneath melts more slowly than nearby, leaving cones and ridges on the snow surface.

References for Chapter 10

Richardson, W.E. (1954). Dirt polygons. Weather 9, 117–121.
Richardson, W.E. & Harper, R.D.M. (1957). Ablation polygons on snow – further observations and theories. Journal of Glaciology 3, 25–27.
Warren Wilson, J. (1954). The initiation of dirt cones on snow. Journal of Glaciology 2, 281–287.
Warren Wilson, J. (1958). Dirt on snow patches. Journal of Ecology 46, 191–198.

AW senior on narrow snow-ridge holds some of the thatch of dead grass at Altan na Beinne

Chapter 11. Photographs showing the use of snow by hill birds and mammals

The photographs mainly show how red grouse and ptarmigan use snow for roosting. A few photographs are of snow holes made by mountain hares, holes that they use as shelter, insulation, and escape cover from predators and human enemies during the day. In one unusual case, a car used for transporting dug peats was left at the peat-banks in early winter and then overwhelmed by snowfall, so it remained all winter and one door was wide open, allowing snow to blow in from outside. Mountain hares made use of the shelter by going inside and digging holes in the snow.

One picture is of a hole where a fox jumped out and ran away when my father and I skied within 5 m of the hole, presumably having heard the noise of the skis and perhaps aware also of vibration through the snow. Inside the hole stretched a conical chamber leading to a circular floor on dry heather, many times larger in area than the remarkably small entrance hole. The fox would have been very sheltered and insulated from the strong wind blowing outside on an exposed hard-packed sheet of snow at about 700 m altitude on a bitterly cold day with no sunshine and thin fog. The site lay high on Meall an Lundain north-east of Derry Lodge near Braemar.

My father took one photograph at 540 m near the old military bridge below Sron na Gaoithe to the east of the road over the Cairnwell pass. An otter had slid down a steep snow bank. Two parallel marks on either side of the main wide central mark from the otter's body indicated that it had glissaded with one foot extended on each side, for stability during the slide and perhaps also for braking. It had then run down at the bottom. Another picture by him shows tracks of an otter along the ice on a hill loch at 685 m altitude above Glen Callater near Braemar.

Snow bowls of a cock and hen ptarmigan close together after overnight roost, the two had formed a close pair for the coming nesting season, Cairnwell, April 1964

Snow burrows of red grouse, small plates of snow show pieces of the roof that fell in when bird rose in morning to put its head through the roof and then walked out, Kerloch, February 1963

As above, ski tracks give scale

Tracks of red grouse to eat heather at dawn after leaving snow burrows, Craiglich, March 1986

Snow bowls of roosting red grouse from overnight, Glen Dye, February 1986

Entry hole of red grouse at dusk is to left of ski stick, exit hole from burrow next morning is further left, Tullochvenus, February 1986

Left of the ski-stick is snow scraped back to fill the entry hole of a burrow of a red grouse when it went to roost the previous evening, Tullochvenus, February 1986

Ptarmigan snow bowls in very icy snow at top of steep Lochnagar cliffs, November 1954

AW stands at snow hole where a fox had just jumped out, Meall an Lundain, early afternoon, February 1963 (Adam Watson senior)

Otter slide on a steep snow bank with ski stick for scale, Sron na Gaoithe, 22 January 1972 (Adam Watson senior)

Otter tracks on Loch Phadruig, 22 November 1970 (Adam Watson senior)

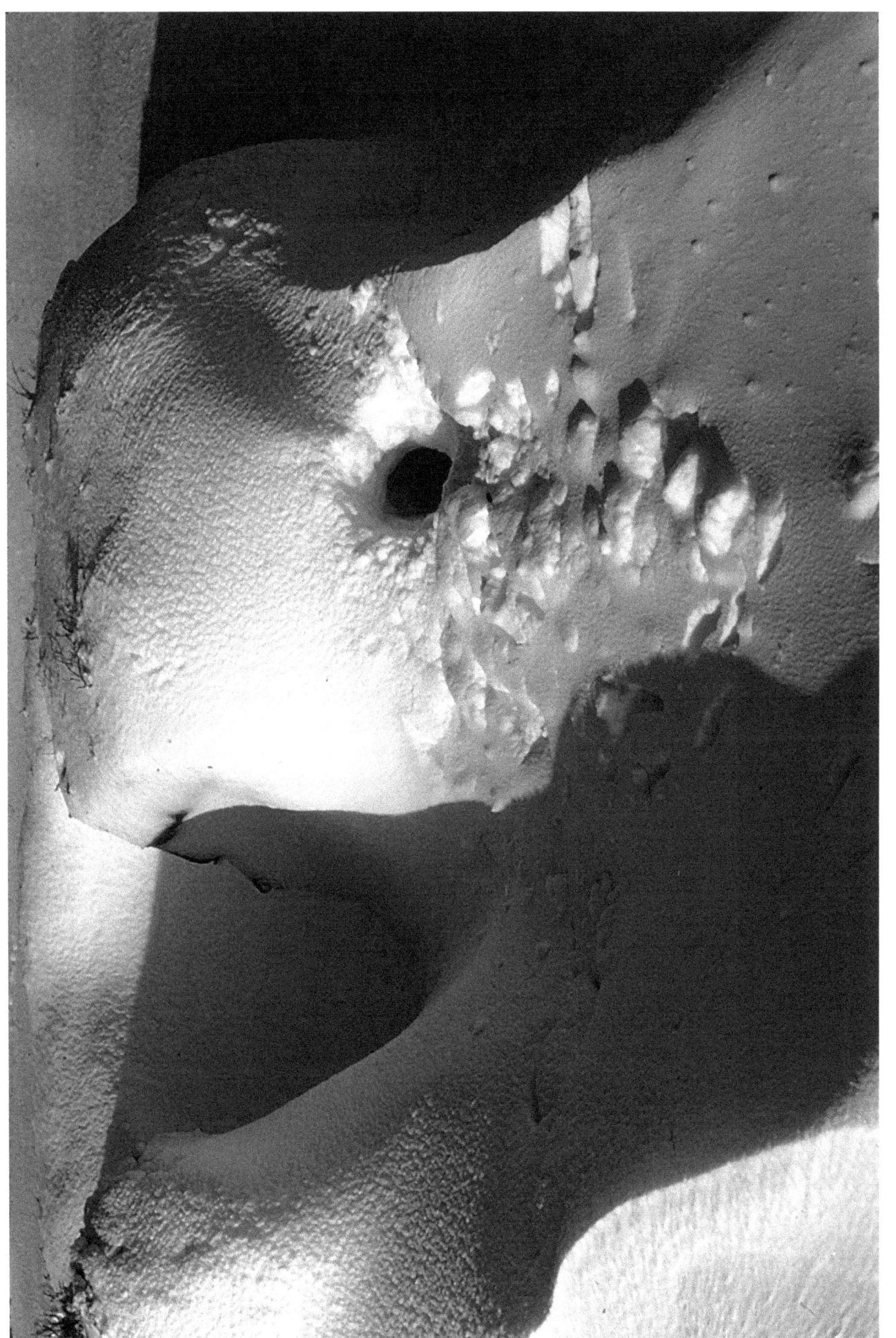

Snow hole of mountain hare at grouse butt, Morven, 2 January 1968 (Adam Watson senior)

A confiding mountain hare stands in sunshine at the entrance to its daytime snow hole in soft powder snow on Carn Dubh in Glen Clunie near Braemar, 10 March 1970 (Adam Watson senior)

Mountain hares use snow in abandoned car for shelter, Glen Gairn, 14 April 1970 (Adam Watson senior)

Mountain hare peeps from daytime snow hole amid hare tracks, Carn Dubh, 10 March 1970 (Adam Watson senior)

Mountain hare peeps from daytime snow hole, Carn Dubh, 10 March 1970 (Adam Watson senior)

Snow pillar with water-eroded grit inside it after a rapid thaw, Coire Domhain, 10 June 1979

Chapter 12. Photographs of some snow features and avalanches

The first photograph shows a snow pillar that contained water-eroded soil inside, after a rapid thaw on snow slopes higher up, leading to a mud slide because thaw water in the upper horizons of soil near and at the surface could not percolate vertically downwards owing to lower horizons being frozen. A second picture shows such a pillar later in the summer, after the snow within it has vanished, leaving the soil contents to collapse and extend outwards in a flat sheet covering the vegetation. The next photograph shows snow holes dug by army personnel for staying overnight, with a close-up revealing dirt and rubbish at the entry holes; the dirt included human faeces. Such holes are likely to reduce the longevity of these semi-permanent snow patches.

Several pictures illustrate avalanches in the Cairngorms, the first two from the annual avalanche site at the Great Slab in Coire an Lochain of Cairn Gorm. A third photograph is of an avalanche in cold midwinter conditions on a steep slope at Cairn of Claise, where freshly blown snow, white of hue in the photograph, has lain on underlying old frozen snow, obvious from its grey colour. This would have been a highly unstable substrate for the fresh snow above it. The fourth photograph is of much interest because the slope gradient is so gentle. However, blowing snow had formed a convex wreath where the upper part had a steeper gradient than the hill slope above, below, and on both sides. Within the avalanche, the snow blocks were sufficiently large as to have posed a serious risk, if not fatal, to anyone in the middle of the site when the avalanche moved. Several photographs show an avalanche in mid May, that had just occurred within a day or two of the photographs being taken. A large convex wreath at the head of Coire na Lairige in Glen Clunie had built up on mat grass *(Nardus stricta),* visible as paler in hue than the dark heather above and below on ground with less snow-lie. The snow had flattened the mat grass, thus providing a slippery surface that was also less stable because of the water from the thawing snow above. The entire snow wreath had slid off from the apex in the centre of the corrie head-wall where the wreath had been deepest. Large blocks of water-saturated heavy snow had then rolled down the slope below, tearing up vegetation and earth. Again, anyone on that slope at the time would have been likely to be killed. The last two photographs are of red deer that were killed in Glen Ey by an avalanche in mid May. The stags had been on a patch of smooth grass showing bright green fresh growth, doubtless attracted by its grazing quality being superior to the dark heather *(Calluna vulgaris)* predominating on the stony slope. A bigger avalanche in Glen Clova killed 37 red deer on 25 or 26 February 1947. At Dog Hillock near Moulzie, it travelled at least 1000 feet, swept across more than 50 yards width of the steep hillside, and attained a breadth of 150 yards where it reached the North Esk (Cairngorm Club Journal 1947, Vol. 16, p.81).

A different snow pillar had collapsed as the snow inside it melted, leaving the grit spread out on the vegetation, Coire Domhain, July 1981

Snow holes by army visitors in Coire Domhain, May 1975

Avalanche on the Great Slab at Coire an Lochain of Cairn Gorm, 21 June 1977

Below the Great Slab on 19 June 1969, avalanche snow almost down to the loch

Avalanche on Cairn of Claise, 14 February 1970

Avalanche on shallow gradient at Sron na Gaoithe, 28 February 1968

Coire na Lairige avalanche, rucksack for scale, note dirt on horizontal layers of snow from different big snowfalls over the winter

Coire na Lairige avalanche, snow blocks have torn up vegetation and earth

Avalanche at Coire na Lairige, Glen Clunie, 18 May 1973

Stag killed by avalanche in Glen Ey, 24 June 1973, avalanche in mid May

Chapter 13. Lichen and moss as indicators of snow lie on cliffs, boulders, soil

Summary

The extent and species of lichen and moss on cliffs, boulders and soil signify the extent of snow-lie. These plants are absent on sites where snow lies very late, or where frequent avalanches down the cliff or water flowing down it prevent the plants from growing.

Introduction

The writer first became aware of this subject in 1951 while climbing in central and north Norway. Cliffs immediately above glacier ice or very long-lying snow patches carried no rock lichens. As a result, the bedrock had the colour and appearance of unweathered fresh rock exposed by rockfalls or by excavation in quarries. The late Mac Smith (1961) was, as far as I know, the first to publish a descriptive account of this in the Cairngorms. Of the rock in the cliffs of at Garbh Choire Mor, the site of the UK's most perennial snow patch, he wrote:- 'It bears a patina of lichen which apparently thrives on granite covered for long periods by snow and is responsible for the remarkable, greenish hue of the buttresses which is most pronounced when they are seen in contrast with the snowfield and gullies.' This was an outstanding observation by Smith, especially if one considers the number of previous writers who had visited the corrie. In fact he had made the observation years before the 1961 publication, and discussed it with me in 1954. Another good observation by him was at Coire Sputan Dearg, where he wrote (1961) of Grey Man's Crag:- 'The basal rocks due to prolonged snow-cover are unweathered and form a curious, pink hem to the brown, upper rocks'. This again he had noticed years earlier and discussed with me in 1954. That was the year when I photographed it and began to make observations on it across the Cairngorms and Lochnagar.

For decades, botanists interested in the species of lichens and mosses that occur in sites with long snow-lie have published accounts with species lists and information on habitat, species abundance and plant communities (for details, see recent reviews in Shaw & Thompson 2006). However, I am unaware of any general descriptive account of species abundance in relation to snow-lie on different parts of cliffs, boulders and soil. The present chapter may therefore be of some use, as an exploratory basis to help stimulate future study in more depth and covering more hill-ranges of the Scottish Highlands.

The main purpose of this chapter is to describe and illustrate major general differences in the colour of cliffs, boulders and soil in the Cairngorms and Lochnagar, associated with the extent and species composition of lichens and mosses, and in turn with variations in snow-lie and certain disturbing factors including avalanche, water flow, and disturbance by human feet.

Descriptions

Smith (1961) noticed that the granite at the very foot of Grey Man's Crag in Coire Sputan Dearg was 'unweathered' and pink in hue (pink is the colour of virgin Cairngorms granite as seen in fresh rockfalls). However, although he noted the unusual greenish hue of the cliffs of Garbh Choire Mor as a result of lichens that favour prolonged snow-lie, he evidently did not notice that the foot of the cliffs there above the most long-lying snow patches was lichen-free and pink, like the foot of Grey Man's Crag. It should be noted that the snow in Garbh Choire Mor is so deep that the pink, lichen-free bedrock is exposed in most years for a very short time in late autumn. Also, Smith spent much more time in Coire Sputan Dearg than in Garbh Choire Mor, and indeed Coire Sputan Dearg was one of his most favoured corries.

In September 2003, after all the snow in Garbh Choire Mor had vanished, David Duncan at my request estimated the height of the lichen-free granite on the cliffs above the main persistent snow patches in Garbh Choire Mor. He

Four stags killed by avalanche in Glen Ey, 24 June 1973, avalanche mid May

judged that it extended about 20 feet up the slab behind the snow site under the foot of Sphinx Ridge, and less behind the snow site under Pinnacles Buttress, which is less persistent than the snow under Sphinx Ridge. At the third most persistent site under the rock climb Michaelmas Fare, he reported that the pale band was present but described it as 'hardly any'. Also he measured by tape the areas without lichen or moss or any other vegetation in the slight hollows under the two most long-lying snow patches. The vegetation-free hollow at the Pinnacles site extended for up to 29 m in length and up to 14 m in width, whereas at the Sphinx site it was smaller, up to 12m x 10m. He noticed that the vegetation-free areas of stones and soil were very loose and unconsolidated, and the bottom of the hollow at Sphinx contained much ptarmigan dung. Also he reported that the vertical depth of the hollows lying below the snow at both the Pinnacles and Sphinx sites reached a maximum of about 60 cm in the centre of each hollow.

At Coire Sputan Dearg, the greenish band even in 1954 occupied a minor part of the lower half of the cliff, most of the rock-face being dark in hue because of rock-lichens that are dark in hue, often almost black. Garbh Choire Mor differed in that all of the cliff was and is greenish, apart from the pink band at the very foot. When David Duncan at my request brought back from Garbh Choire Mor a small piece of granite with greenish rock-lichens in the 2000s, the main lichen identified by Dr Brian J. Coppins of Edinburgh (for more detail see Appendix 3 of Chapter 4) was *Rhizocarpon geographicum,* which is greenish-yellow in colour.

In 1954 I noticed several other features of the colour of cliffs in Coire Sputan Dearg, and this has been confirmed in many photographs taken in the last three years. Where the gradient of the cliff is relatively low, as at Crystal Ridge, the greenish band extends further up than usual for the corrie. Where the cliff is steep, as at Black Tower, there is no obvious greenish band. Blowing and falling snow piles up to a greater height on top of the rock on the easy-gradient cliffs, whereas on steep cliffs it falls to the bottom and does not pile up there unless there is a flat platform at the foot. Blowing snow past the cliff-top piles up in easy-angled gullies and bedrock, and there one finds the greenish band of rock at the cliff-top and also down the gully and on any easy-angled bedrock on either side.

At the longest-lying snow patches at Garbh Choire Mor, the lichen-free pink band on bedrock along the cliff-foot also extends to snow-patch hollows with boulders and soil below the cliff. At a few metres' distance away from the edge of the snow-patch hollow, tiny clumps of dark green moss species that can withstand long snow-lie grow on boulders and soil, as well as in moist cracks in the granite bedrock along the lichen-free band above. This can also be seen at other sites with snow that is less long-lying, including Garbh Uisge Beag, hollows in lower Garbh Uisge Mor, and Ciste Mhearad of Cairn Gorm. It is clear that moss can grow on bedrock and boulders nearer to the longer-lying snow patches than can be attained by rock-lichens. One difference, however, is that such mosses are usually dotted as small clumps that are scattered and well dispersed on bedrock and boulders, whereas rock-lichens that have colonised bedrock and boulders generally cover the surface to a far greater extent, obscuring the colour of the bedrock or boulders underneath.

Photographs on Beinn a' Bhuird at Polypody Groove and Mitre Ridge show that where the lower part of a cliff has a relatively small gradient, a large proportion of the cliff, almost a half of it, has the pale hue from greenish rock-lichens, even though the build-up of snow is far less than at Garbh Choire Mor and typically all snow has gone by late summer on the cliff and below it. Boulders receiving much snow-lie beside cliff-tops and on ground immediately at the foot of cliffs show more of the greenish lichen than boulders on more exposed ground away from the immediate cliff-top or cliff-foot.

Where frequent avalanches occur, as on the Great Slab of Coire an Lochain on Cairn Gorm, the granite bedrock remains in the virgin pink colour because of the tearing action of snow and rocks preventing lichens and mosses from becoming established. In such sites, the only vegetation in material amount consists of moss in moist or wet cracks that are not exposed to the avalanches. At the Great Slab, groundwater also flows down the bedrock, which adds to hostile conditions for lichens and mosses trying to colonise rock. At some cliffs such as Stac an Fharaidh, groundwater springs issuing above the cliff-top keep the bedrock free from vegetation.

The greenish-yellow lichen-band occurs on snowy sections on all the main cliffs in the Cairngorms massif, Lochnagar and the White Mounth. Where parts of a cliff or ground at a path along the cliff-top are heavily trampled by climbers or walkers, rock-lichens tend to be worn off, exposing more of the underlying pink colour of the granite.

On the cliffs of Ben Nevis the bedrock is mainly granite, though with andesitic lava on the highest sections, and

Garbh Choire Mor, green chionophilous lichens on the cliffs apart from the pink lichen-free cliff-foot and lichen-free boulders beside the sites of the longest-lying snow patches, which had all melted in early September, 7 October 1959 (Alex Tewnion)

Garbh Choire Mor below Sphinx Ridge after all snow had gone in early September, showing lichen-free lower cliff, boulders and soil, and a layer of black dirt under the site of the longest-lying snow, 7 October 1959 (Alex Tewnion)

Garbh Choire Mor below Sphinx Ridge, as in previous photo, 7 October 1959 (Alex Tewnion)

Garbh Choire Mor below Pinnacles Buttress after all snow had gone, as in previous three photos and showing moss as the vegetation nearest the snow patch, 13 September 1959 (Adam Watson senior)

so some different species of rock-lichen would be expected as well as different proportions of them. To judge from photographs at the long-lying snow patch in Observatory Gully, the bedrock and boulders beside the remaining snow in autumn appear to lack rock-lichens, as at Garbh Choire Mor. However, bedrock further from the snow patch carries more chionophilous dark green moss than in the Cairngorms, associated with the much heavier rainfall and more persistent fog at Ben Nevis and also the greater shade at the snow patch in Observatory Gully even when the sun is shining. Also, such moss abounds on boulders much nearer the snow patch than at Garbh Choire Mor. It should be noted that similar comments can be made about lichen-free bedrock and moss beside the long-lasting snow patch at the foot of Point Five Gully on the Ben Nevis cliffs, and beside other persistent snow patches on the cliffs of Ben Nevis and the nearby Aonach Mor (granite) and Aonach Beag (schist).

Measurements of the extent of the pale band of rock-lichens in different years should show whether the extent has altered in relation to climatic changes, and also the rate of change. Likewise, the extent of the lichen-free band of bedrock, boulders and soil at the longest-lying snow patches, and any changes in it over time, should be associated with climatic changes. Any reduction in its extent on boulders and soil would be expected to be matched by increases of moss from the moss-dominated less snowy ground nearby.

Further reading and viewing

Readers will see detailed descriptions and discussions on lichens in relation to snow-lie within the website www.winterhighland.info, thread *Attention all hill walkers! Snow patch season 2009*, in June 2009, and then from 25 June onwards in a new thread *Lichenometry etc*, started by 'Firefly' (Iain Cameron). In both threads, especially the second one, photographs show the pale, greenish band of rock-lichens at various cliffs in the Cairngorms area, and a pale lichen-band at snow patches in the north-west Highlands as well as on Ben Nevis and nearby Aonach Mor. In particular, 'Scomuir' (Scott Muir) took many new photographs in 2009 while searching for the lichen band on cliffs across the Cairngorms and Lochnagar, and contributed valuable comments and discussions.

Acknowledgements

I thank David Duncan and Dr Brian J. Coppins for useful information.

References

Shaw, P. & Thompson, D.B.A. (Eds) (2006). The nature of the Cairngorms. The Stationery Office, Edinburgh.
Smith, M. (1961). Climbers' guide to the Cairngorms area. Scottish Mountaineering Club, Edinburgh.

A few published photographs
These illustrate the pale band of rock-lichens on cliffs at persistent snow patches in the Cairngorms and Lochnagar
Alexander, H. (1928). The Cairngorms. Scottish Mountaineering Club District Guide. SMC, Edinburgh.
Opposite p.170. Coire an Dubh-lochain, pale rock at the cliff-foot above the centre of the loch.
1938 edition
Opposite p171. lower photograph, Mitre Ridge, wide band of pale rock on the lower part of Mitre Ridge (prominent steep ridge right of centre) and on the cliff-foot up the gully to the left.
1950 edition
Opposite p175. above snow patch at the foot of Shelter Stone Crag.
Opposite p182. ditto, and also low rocks above snow patch to the right of Garbh Uisge.
Opposite p190. Coire Sputan Dearg, conspicuous rock-fall at the top of Crystal Ridge on its south side, pale band of rock-lichen at foot of much of the cliff and far up the gullies beside the cliff, e.g. conspicuous on Pinnacles Buttress, Crystal Ridge, No 2 Buttress (now called Grey Man's Crag), Anchor Route, Cherub's Buttress.
Opposite p207. Coire Bhrochain. Pale band of rock-lichen conspicuous at foot of B, C and D.

Coire Sputan Dearg, showing pale lichen band on the cliff-foot above the snow patches and extending up gullies filled with snow, in places to the cliff-top, whereas steep cliffs that carry far less snow support dark rock-lichen, almost black, 5 June 1954

After p238, third photo in the set, two climbers looking at cliff in Coire an Dubh-lochain, wide band of pale rock-lichen goes almost half way up the cliff at Polypody Groove.

Watson, A. (1975). The Cairngorms. SMC District Guide. Scottish Mountaineering Trust, Edinburgh.

Plate 30, Coire Bhrochain, along the foot of the cliffs and beside higher gullies that often hold snow.

Gimingham, C. (Ed) (2002). The ecology, land use and conservation of the Cairngorms. Packard Publishing, Chichester.

Plate 13. Coire nan Clach of Beinn a' Bhuird, pale rock conspicuous at the cliff-foot on the sunlit right part of photograph.

Strange, G. (2010), The Cairngorms: 100 years of mountaineering. Scottish Mountaineering Trust.

p32. Mitre Ridge from the west, shows pale band above the main part of the snow patch.

pp.72, 73. Coire an Dubh-lochain, Polypody Groove, wide pale band on the lower cliff which is less steep and so carries snow for a longer period. Smith (1961) wrote 'Winter: The 300-foot lower section is completely masked by a snowfield; Polypody Groove as such does not exist'.

p191. Garbh Choire Mor, lichen free pink granite is obvious at the bottom left on the cliff-foot. Above it stretches a wide pale band of greenish rock-lichen, except on the steepest rock.

Chapter 14. Some photographs of snow mould on hill vegetation

It is well known that snow moulds, commonly species of the genus *Fomes* or *Fusarium* grow on grass in lowland towns and countryside, and can kill the grass in localised patches. Because of the longer snow-lie at higher altitudes and in more northerly latitudes, snow mould is more prevalent and grows on a wide variety of plant species. Most species of plant in the alpine zone of Scotland, in subarctic regions such as north Iceland and northernmost Norway, and in Arctic regions such as in north Canada are resistant to the mould. Even though the mould can grow on many plant species there, it does not kill a material amount of foliage. However, heather or ling (*Calluna vulgaris*) is particularly susceptible and the foliage can easily be killed, with the result that very old plants in the degenerate phase of growth can be killed completely. Heather does not occur in the snowy regions of Arctic Canada such as east Baffin Island, and perhaps one reason for this is that it could not withstand the long snow-lie and the snow moulds.

The moulds cannot grow when the snow and vegetation are frozen, but appear when a thaw exposes the vegetation just outside the snow patch and when the air temperature is above freezing. Plants there are wet from the recent snow melting, and tall old heather tends to be easily flattened into a sheet by the weight of the snow. The wet sheet does not dry readily and creates good conditions for the mould to spread rapidly. Usually the mould is white or grey, but occasionally dark grey verging on black, especially when growing on hill grasses in spots that are wet from temporary water-logging during snow-melt. The whitest patches look almost like a dusting of snow, and reflect sunlight.

Arnthor Garðarsson kneels at heather affected by snow mould at Hrisey, May 1965

Snow patch at Hrisey, north Iceland, showing a film of pale snow mould growing on grass beside the snow, May 1965

Real snow on right, snow mould on left on grasses at Cairn Lochan, 30 May 1982

Snow mould growing on grasses at Cairn Lochan, 30 May 1982

Latest information

The cooling trend of recent years was maintained in May and June 2011, with many days of fresh snowfall. This reached a record 9 days in June 2011 on the high Cairngorms, a maximum value attained only once previously, in 1977. Following the unusually heavy snowfalls and cold of November 2010 to February 2011 in the Highlands, I expected that the extent of consolidated snow cover on Ben Macdui plateau at the start of June would be large, as in 2010. However, a March with relatively few snowfalls, an April of record warmth, and a May with frequent heavy rain put paid to that. In the end, the proportion of the plateau covered by consolidated snow was only 10%.

Some other books by the author

1963. Mountain hares. Sunday Times Publications, London (with R. Hewson)

1970. Editor of Animal populations in relation to their food resources. Symposia of the British Ecological Society No 10, published by Blackwell Scientific Publications, Oxford and Edinburgh.

1974. The Cairngorms, their natural history and scenery. Collins, London, and 1981 Melven Press, Perth (with D. Nethersole-Thompson)

1975. The Cairngorms. Scottish Mountaineering Club District Guide, published by the Scottish Mountaineering Trust, second edition published 1992

1976. Grouse management. The Game Conservancy, Fordingbridge, and the Institute of Terrestrial Ecology, Huntingdon (with G.R. Miller)

1982. Animal population dynamics. Chapman and Hall, London and New York (with R. Moss and J. Ollason)

1982. The future of the Cairngorms. The North East Mountain Trust, Aberdeen (with K. Curry-Lindahl and D. Watson)

1984. The place names of upper Deeside. Aberdeen University Press, Aberdeen (with E. Allan)

1998. The Cairngorms of Scotland. Eagle Crag, Aberdeen (with S. Rae)

2008. Grouse, the grouse species of Britain and Ireland. HarperCollins, London, Collins New Naturalist Library No 107 (with R. Moss)

2010. Cool Britannia, snowier times in 1580–1930 than since. Paragon Publishing, Rothersthorpe, Northants (with I. Cameron)

2011. It's a fine day for the hill. Paragon Publishing, Rothersthorpe, Northants

2011. A zoologist on Baffin Island, 1953. Paragon Publishing, Rothersthorpe, Northants

2011. Vehicle hill tracks in northern Scotland. Published by North East Mountain Trust, Aberdeen. Published imprint Paragon Publishing, Rothersthorpe, Northants.

www.ingramcontent.com/pod-product-compliance
Lightning Source LLC
Chambersburg PA
CBHW041411300426
44114CB00028B/2983